A textbook

on

Applied Integral Equations

Ramakanta Meher

ISBN:978-83-66675-56-8 - print
ISBN:978-83-66675-57-5 - online

https://doi.org/ 10.2478/9788366675575-fm

Preface

There is a vital role of differential and integral equations in studying different types of real-world problems to study the behaviour of the issues. Thus, it becomes essential to know the various methods of finding solutions of the integral equation in explicit form. For the integral equations whose solutions cannot be found in explicit form, one has to study the properties of solutions of the given differential equation to guess an approximate solution.

This textbook entitled ***Applied Integral Equations*** is intended to study the methods of finding the explicit solutions of integral equations where ever possible and in the absence of finding an exact solution. It is intended to study the properties of solutions of the given integral equations. This book contains 07 chapters. *Chapter-1* discusses the introduction to integral equations, classification of integral equations, Relation between linear differential equations and Volterra integral equation, Nonlinear equation and solution of an integral equation. *Chapter-2* discusses the existence and uniqueness theorems of Integral equations, Successive approximation, Iterated Functions, Reciprocal functions, Volterra Solution of Fredholm's equation, Discontinuous Solution, Fredholm equations with separable kernels and

Resolvent Kernel. *Chapter-3* discusses the Fredholm equation as a limit of a finite system of linear equations, Hadamard's Theorem, Fredholm's two fundamental relations, Fredholm's solution of the Integral equation for different $D(\lambda)$, Characteristic numbers and fundamental functions, the associated Homogenous integral equations, the orthogonality theorem, Kernels of the form $\sum a_i(x)b_i(y)$, Eigen Values and eigenfunctions, Fredholm integral equation of the second kind, Eigenvalues for non-separable kernels, Volterra Integral Equation, Solution by the Resolvent kernel and Method of successive approximation. *Chapter-4* discusses the Applications of Fredholm theory, Free vibration of an elastic string, The differential equation of the problem, Reduction to a dimensional BVP, Solution of the boundary value problem, Construction of Green function, Equivalence between the Boundary value problem and Linear integral equations, Constrained vibrations of an elastic String, Equivalence between boundary value problem and Linear integral equations and Remark on the solution of the BVP. *Chapter-5* discusses the Hilbert-Schmidt Theory that includes Iterations of symmetric kernels, Orthogonality theorem, An existence theorem for the nonlinear integral equation of Fredholm type and the equation of Bratu. *Chapter-6* discusses the Fredholm alternatives, An example of Picard's method, Powers of an integral operator, Iterated kernels, Neumann series, A remark on the convergence of the iterative method, Differentiation of function under an integral sign, Relation between differential and integral equation, The Fredholm alternatives and the Fredholm alternative theorem.*Chapter-7* discusses the method of undetermined coefficients that includes approximation methods of undetermined coefficients, the method of collocation, the method of weighting functions, the method of least squares and approximation of the kernel.

This book is based on syllabi of the theory of integral equations prescribed for the undergraduate and postgraduate students of mathematics and PhD students in different institutions and universities of India and abroad. This book will be helpful for the competitive examinations as well.

I shall heartily welcome constructive criticisms, views and suggestions from the reviewers and readers to further improve the book.

Ramakanta Meher

https://doi.org/ 10.2478/9788366675575-toc

Contents

https://doi.org/ 10.2478/9788366675575-toc

https://doi.org/ 10.2478/9788366675575-toc

https://doi.org/ 10.2478/9788366675575-toc

https://doi.org/ 10.2478/9788366675575-toc

https://doi.org/ 10.2478/9788366675575-001

Chapter-1

Integral equation

1.1. Introduction:

In linear Fredholm and Volterra integral equation of the first kind, the unknown function $y(x)$ appears linearly only under the integral sign. In contrast, it appears linearly inside the integral sign and outside the integral signals in the second kind of linear Fredholm and Volterra Integral equations.

1.2. The Limit of integration:

The integral is over a finite interval with a fixed limit of integration in the Fredholm Integral equation. However, in the Volterra Integral equation, at least one Limit of the range of integration is a variable upper limit is the most commonly used with a variable limit.

1.3. Singular Behaviour of the Integral equations:

If the integration is improper in an integral equation, then it is called a *singular integral equation*. Singular integral equation usually occurs if the interval of integration is infinite or if the kernel becomes unbounded at one or more points of the interval of consideration i.e., $a \le t \le b$.

1.4. Integro differential equation:

In 1900, Volterra studied the population growth where new types of equations developed, termed as Integro differential equation.

In this equation, the unknown function occurs on one side as an ordinary derivative and appears on the other side under the integral sign.

For Example:

$$\frac{dy}{dx} = -\cos x - 1 + \int_0^x y(t)\,dt \;, y(0) = 1 \tag{1.4.1}$$

$$\frac{d^2y}{dx^2} = -x^2 + \int_0^x (x-t)\,y(t)\,dt \;, y(0) = 0, y'(0) = 1 \tag{1.4.2}$$

$$\frac{dy}{dx} = 1 - \frac{2}{3}x + \int_0^x xty(t)\,dt \;, y(0) = 1 \tag{1.4.3}$$

Equation (1.4.1) and (1.4.2) are *Volterra Integro differential* equations, and (1.4.3) is a *Fredholm Integro-differential* equation.

https://doi.org/ 10.2478/9788366675575-001

1.5. Singular Integral Equation:

The integral equation of the first kind

$$f(x)=\lambda\int_{\alpha_1(x)}^{\alpha_2(x)} K(x,t)y(t)dt \tag{1.5.1}$$

and the Integral equation of the second kind

$$y(x)=f(x)+\lambda\int_{\alpha_1(x)}^{\alpha_2(x)} K(x,t)y(t)dt \tag{1.5.2}$$

are called *Singular*, if the lower limit, the upper limit, or both limits of integration are infinite.

Similarly, equation (1.5.1) and (1.5.2) are also called *singular integral equation* if the kernel $K(x,t)$ becomes infinite at one or more points in the integration domain.

For example:

The following examples describe the singular integral equation of the first type

$$y(x)=x+5\int_{0}^{\infty} Cos(x-t)y(t)dt \tag{1.5.3}$$

$$y(x)=x^2+\frac{2}{3}\int_{-\infty}^{0} Sin(x+t)y(t)dt \tag{1.5.4}$$

$$y(x)=1+x+\frac{5}{6}\int_{-\infty}^{\infty} (x+t)y(t)dt \tag{1.5.5}$$

The range of integration, i.e. the lower limit or the upper limit of the integration, is infinite.

Examples of the second kind of singular integral equations are given by

$$x^2 = \int_0^x \frac{1}{\sqrt{x-t}} y(t)\,dt \tag{1.5.6}$$

$$x = \int_0^x \frac{1}{(x-t)^\alpha} y(t)\,dt \ , 0 < \alpha < 1 \tag{1.5.7}$$

$$y(x) = 1 + \sqrt{x} - \int_0^x \frac{1}{\sqrt{x-t}} y(t)\,dt \tag{1.5.8}$$

The singular behaviour in this kind of equation has resulted from the kernel becoming infinite $t \to x$.

An integral equation is similar to *Abel's integral equation* and generalized *Abel's Integral equations*.

These singular integral equations are established by the Norwegian mathematician Niel's Abel in 1823, which are among the earliest integral equation.

Singular integral equations similar to type (1.5.8) are called the *weakly-singular second kind Volterra type integral equations*. These equations usually arise in many science and engineering applications like superfluidity problems, Heat conduction, and crystal growth.

1.6. Volterra-Fredholm Integral Equation:

Volterra-Fredholm Integral Equation is a disjoint combination of Volterra and Fredholm integrals that appears in one Integral equation.

These integral equations arise from the Modelling of the spatiotemporal development of an epidemic from boundary value problems and many biological and chemical applications.

The standard form of the *Volterra –Fredholm integral equation* can be expressed as

$$y(x)=f(x)+\int_0^x K_1(x,t)y(t)dt+\int_a^b K_2(x,t)y(t)dt \qquad (1.6.1)$$

Where $K_1(x,t)$ and $K_2(x,t)$ are the kernels of the equations.

For example:

The Volterra –Fredholm integral equation is

$$y(x)=2x-\int_0^x (x-t)y(t)dt+\int_0^{\frac{\pi}{2}} xy(t)dt \qquad (1.6.2)$$

https://doi.org/ 10.2478/9788366675575-001

and $y(x)=\sin x-\cos x-\int_0^x y(t)\,dt+\int_0^{\frac{\pi}{2}} y(t)\,dt$ (1.6.3)

Here, the unknown function $y(x)$ appears inside the Volterra and Fredholm integrals and outside both integrals.

1.7. Volterra-Fredholm Integro-Differential Equations:

A combination of disjoint Volterra and Fredholm integrals and differential operators is called the Volterra-Fredholm Integro-differential equation, and it may appear in one Integral equation.

These types of integral equation arise in many physical and chemical applications similar to the Volterra-Fredholm integral equations.

The standard form of the Volterra –Fredholm integro-differential equation can be expressed as

$$\frac{d^n y}{dx^n}=f(x)+\int_0^x K_1(x,t)y(t)\,dt+\int_a^b K_2(x,t)y(t)\,dt \qquad (1.7.1)$$

Where $K_1(x,t)$ and $K_2(x,t)$ are the kernels of the equation and is the order of the ordinary derivative of $y(x)$.

It can be noticed that since this kind of equations contains ordinary derivatives, so the initial conditions should be prescribed depending upon the order of the derivatives involved.

Example-1

The Volterra-Fredholm Integro differential equations are

$$\frac{dy}{dx}=1+\int_0^x (x-t)\,y(t)\,dt+\int_0^1 xty(t)\,dt\ , \qquad (1.7.2)$$
$$y(0)=1$$

and

$$\frac{d^2y}{dx^2}=-x-\frac{1}{6}x^3+\int_0^x y(t)\,dt+\int_{-\pi}^{\pi} xy(t)\,dt,$$
$$y(0)=0, y'(0)=2$$

Here the unknown function $y(x)$ appears inside the Volterra and Fredholm integrals and outside both integrals.

Ordinary differential equations and Volterra Integral equation:

Consider the equation

$$\frac{d^ny}{dx^n}+a_1(x)\frac{d^{n-1}y}{dx^{n-1}}+a_2(x)\frac{d^{n-2}y}{dx^{n-2}}+.....+a_n(x)\,y \qquad (1.9.1)$$
$$=\phi(x)$$

Where the origin is a regular point for the $a_i(x)$.

Let us make the transformation

$$\frac{d^n y}{dx^n} = z(x)$$

Then

$$\frac{d^{n-1}y}{dx^{n-1}} = \int_0^x z(x)\,dx + C_1$$

$$y = \int_0^x z(x)\,dx^n + C_1 \frac{x^{n-1}}{(n-1)!} + C_2 \frac{x^{n-2}}{(n-2)!} + \ldots + C_n \qquad (1.9.2)$$

Where $\int_0^x z(x)\,dx^n$ stands for a multiple integral of order n.

Equation (1.9.2) transform (1.9.1) in to

$$z(x) + a_1(x)\int_0^x z(x)\,dx + \ldots\ldots + a_n(x)\int_0^x z(x)\,dx^n = \phi(x) + \sum_{i=1}^{n} C_i \alpha_i(x) \qquad (1.9.3)$$

Where $\alpha_i(x) = a_i(x) + \frac{x}{1} a_{i+1}(x) + \ldots\ldots + \frac{x^{n-i}}{(n-i)!} a_n(x)$.

If we now put

$$\phi(x) + \sum_{i=1}^{n} C_i \alpha_i(x) = f(x)$$

And make use of the well-known formula

$$\int_0^x z(t)\,dt^n = \int_0^x \frac{(x-t)^{n-1}}{(n-1)!} z(t)\,dt$$

Equation (1.9.3) becomes

$$z(x) + \int_0^x \left[a_1(x) + a_2(x)(x-t) + \ldots\ldots.. a_n(x) \frac{(x-t)^{n-1}}{(n-1)!} \right] z(t)\,dt$$
$$= f(x)$$

Which is a Volterra Integral equation of the second kind.

1.8. Nonlinear equation:

In this type of Integral equation, the unknown function may appear in the equation to a power n greater than 1.

For example:

$$y(x) = f(x) + \lambda \int_a^b K(x,t)\, y^n(t)\,dt$$

The unknown function may appear in a more general way, as indicated by the following equation:

$$y(x) = f(x) + \lambda \int_a^b \phi\left[x, t, y(t)\right] dt$$

In particular, the differential equation

$$\frac{dy}{dx} = \phi(x, y)$$

can be put in the integral form

https://doi.org/ 10.2478/9788366675575-001

$$y(x)=C+\int_{0}^{x}\phi\left[t,y(t)\right]dt$$

Still, other general types of nonlinear Integral equations have been considered.

1.9. Singular Integral Equation:

An integral equation is singular when either one or both of the limits of integration become infinite.

For example:

$$y(x)=f(x)+\lambda\int_{0}^{\infty}Sin(xt)\,y(t)\,dt$$

An integral equation is also singular if the kernel becomes infinite for one or more points of the interval under discussion.

For example:

$$f(x)=\int_{0}^{x}\frac{H(x,t)}{(x-t)^{\alpha}}y(t)\,dt\ ,\ (0<\alpha<1)$$

Abel's problem, i.e.,

$f(x)=\int_{a}^{x}\frac{y(t)\,dt}{(x-t)^{\alpha}}\ ,\ (0<\alpha<1)$ is itself a singular Integral equation.

https://doi.org/ 10.2478/9788366675575-001

1.10. Solution of an integral equation:

A solution of an integral equation or an integrodifferential equation on the interval of integration is a function $y(x)$ such that it satisfies the given equation.

The critical concept will be illustrated first by examining the following examples.

Example-2

Show that $y(x)=x$ is a solution of the following Fredholm integro-differential equation.

$$\frac{dy}{dx}=\frac{2}{3}+\int_0^1 t.y(t)\,dt \qquad (1.12.1)$$

Substituting $y(x)=x$ in the right-hand side of (1.12.1), we obtain

$$=\frac{2}{3}+\int_0^1 t^2 dt=\frac{2}{3}+\left[\frac{t^3}{3}\right]_0^1=1=\frac{dy}{dx}$$

Example-3

Show that $y(x)=x$ is a solution of the following Fredholm integral equation

$$y(x)=\frac{5}{6}x-\frac{1}{9}+\frac{1}{3}\int_0^1 (x+t).y(t)\,dt \qquad (1.12.2)$$

Substituting $y(x)=x$ in the Right-hand side of (1.12.2), we obtain

$$=\frac{5}{6}x-\frac{1}{9}+\frac{1}{3}\int_0^1 (x+t).y(t)\,dt$$

$$=\frac{5}{6}x-\frac{1}{9}+\frac{1}{3}\left[\frac{xt^2}{2}+\frac{t^3}{3}\right]_0^1$$

$$=x=y(x)$$

Example-4

Check $y(x)$ is a solution of the following Fredholm integral equation Where $y(x)=x+e^x$.

$$y''(x)=e^x-\frac{4}{3}x+\int_0^1 xt.y(t)\,dt, \qquad (1.12.3)$$

$$y(0)=1, y'(0)=2$$

Substituting $y(x)=x+e^x$ in the right-hand side of (1.12.3) We obtain

$$=e^x-\frac{4}{3}x+\int_0^1 xt.(t+e^t)\,dt=e^x=y''(x)$$

Example-5

https://doi.org/ 10.2478/9788366675575-001

Show that If $y(x)=x^2+x^3$ is a solution of the Fredholm Integral equation

$$y(x)=x^3-x^2-2x+\alpha\int_{-1}^{1}\left(xt^2+x^2t\right).y(t)\,dt \qquad (1.12.4)$$

Then find α.

Substituting $y(x)=x^2+x^3$ into both sides of equation (1.12.4), we obtain

$$x^2+x^3=x^3-x^2-2x+\alpha\left(\frac{2}{5}x^2+\frac{2}{5}x\right)$$

Solving for α, It gives $\alpha=5$. □

https://doi.org/ 10.2478/9788366675575-001

Exercises

Verify if the given functions are solutions of the corresponding integral equations:

1. $$y(x)=e^{x}\left(2x-\frac{2}{3}\right),\quad y(x)=2xe^{x}-2\int_{0}^{1}e^{(x-t)}y(t)\,dt.$$

2. $$y(x)=e^{x},\quad y(x)+\lambda\int_{0}^{1}\sin xt\,y(t)\,dt=1.$$

3. $$y(x)=1-\frac{2\sin x}{1-\frac{\pi}{2}},\quad y(x)=1+\int_{0}^{\pi}\cos(k+t)\,y(t)\,dt.$$

4. $$y(x)=\cos x,\quad y(x)=\sin x+\int_{0}^{\pi}\left(x^{2}+t\right)\cos t\,y(t)\,dt.$$

5. $$y(x)=x-\frac{x^{3}}{6},\quad y(x)=x-\int_{0}^{x}\sinh(x-t)\,y(t)\,dt.$$

https://doi.org/ 10.2478/9788366675575-001

6. $$y(x) = \frac{1}{2},$$
$$\int_0^x \frac{y(t)}{\sqrt{x-t}}\,dt = \sqrt{x}.$$

https://doi.org/ 10.2478/9788366675575-002

Chapter-2

Existence and Uniqueness Theorems of Integral equations

2.1. Fredholm Integral Equation:

Here we consider a Fredholm linear integral equation of the second kind with a parameter.

$$y(x) = f(x) + \lambda \int_a^b K(x,t)\, y(t)\, dt \qquad (2.1.1)$$

We take up the first case where both limits of integration are fixed. (***Fredholm equation***)

Let us assume the following conditions for a Fredholm Integral equation:

(a) $y(x) = f(x) + \lambda \int_a^b K(x,t)\, y(t)\, dt$

(b) $K(x,t) \neq 0$, is real and continuous in the rectangle R, i.e. $\{R : a \leq x \leq b, a \leq t \leq b\}$.

(c) $f(x) \neq 0$ is real and continuous in the interval I for which $a \leq x \leq b$.

(d) λ constants.

https://doi.org/ 10.2478/9788366675575-002

We see at once that if there exists a continuous solution $y(x)$ of (2.1.1) and $K(x,t)$ is continuous, then $f(x)$ must be continuous. Hence the inclusion of condition **(c)** above.

Substitute in the second member of (2.1.1) in place of $y(t)$, it gives

$$y(x)=f(x)+\lambda\int_a^b K(x,t)\left[f(t)+\lambda\int_a^b K(t,t_1)y(t_1)dt_1\right]dt$$

$$=f(x)+\lambda\int_a^b K(x,t)f(t)dt$$

$$+\lambda^2\int_a^b K(x,t)\int_a^b K(t,t_1)y(t_1)dt_1dt$$

Again, substitute for $y(t_1)$ its value as given by (2.1.1).

It finds

$$y(x)=f(x)+\lambda\int_a^b K(x,t)f(t)dt$$

$$+\lambda^2\int_a^b K(x,t)\int_a^b K(t,t_1)\left[f(t_1)+\lambda\int_a^b K(t_1,t_2)y(t_2)dt_2\right]dt_1dt$$

$$=f(x)+\lambda\int_a^b K(x,t)f(t)dt+\lambda^2\int_a^b K(x,t)\int_a^b K(t,t_1)f(t_1)dt_1dt$$

$$+\lambda^3\int_a^b K(x,t)\int_a^b K(t,t_1)\int_a^b K(t_1,t_2)y(t_2)dt_2dt_1dt$$

Proceeding in this way, it obtains

https://doi.org/ 10.2478/9788366675575-002

$$y(x)=f(x)+\lambda\int_a^b K(x,t)f(t)\,dt$$
$$+\lambda^2\int_a^b K(x,t)\int_a^b K(t,t_1)f(t_1)\,dt_1dt+.....$$
$$+\lambda^n\int_a^b K(x,t)\int_a^b K(t,t_1).....\tag{2.1.2}$$
$$..\int_a^b K(t_{n-2},t_{n-1})f(t_{n-1})\,dt_{n-1}...dt_1dt+R_{n+1}(x)$$

Where

$$R_{n+1}(x)=\lambda^{n+1}\int_a^b K(x,t)\int_a^b K(t,t_1).....\int_a^b K(t_{n-1},t_n)y(t_n)\,dt_n...dt_1dt$$

This leads us to the consideration of the following infinite series:

$$f(x)+\lambda\int_a^b K(x,t)f(t)\,dt$$
$$+\lambda^2\int_a^b K(x,t)\int_a^b K(t,t_1)f(t_1)\,dt_1dt+......\tag{2.1.3}$$

Under our hypothesis **(b)** and **(c)**, each term of this series is continuous in I. This series then represents a constant function in I, provided it converges uniformly in I .

Since $K(x,t)$ and $f(x)$ are continuous in R and I respectively, $|K|$ has a maximum value M in R and $|f(x)|$ has a maximum value N in I:

$|K(x,t)| \le M$ in R.

$|f(x)| \le N$ in I.

Putting

$$S_n(x) = \lambda^n \int_a^b K(x,t) \int_a^b K(t,t_1) \ldots\ldots \int_a^b K(t_{n-2},t_{n-1}) f(t_{n-1}) dt_{n-1} \ldots dt_1 dt$$

Then $|S_n(x)| \le |\lambda^n| NM^n (b-a)^n$

The series of which this is a general term converges only when

$|\lambda| M(b-a) < 1$

Thus, we see that the series (2.1.3) converges absolutely and uniformly, when

$$|\lambda| < \frac{1}{M(b-a)}.$$

https://doi.org/ 10.2478/9788366675575-002

If (2.1.1) has a continuous solution, then it must be expressed by (2.1.2). If $y(x)$ is continuous in I and its absolute value has a maximum value Y.

Then

$$|R_{n+1}(x)| < |\lambda^{n+1}| Y M^{n+1} (b-a)^{n+1}$$

If $|\lambda| M(b-a) < 1$,

then $\lim_{n\to\infty} R_{n+1}(x) = 0$.

Thus, it can be seen that the function $y(x)$ satisfying equation (2.1.2) is the continuous function that is given by the series (2.1.3).

It can be verified by direct substitution that the function $y(x)$ defined by (2.1.3) satisfies (2.1.1) or what amounts to the same thing, place the series given by (2.1.3) equal to $y(x)$, multiply both sides by $\lambda K(x,t)$ and integrate term by term, as we have a right to do.

We obtain

$$\begin{aligned}\lambda\int_a^b K(x,t)y(t)dt &= \lambda\int_a^b K(x,t)\left[f(t)+\lambda^2\int_a^b K(t,t_1)f(t_1)dt_1+....\right]dt\\ &= \lambda\int_a^b K(x,t)f(t)dt+\lambda^2\int_a^b K(x,t)\int_a^b K(t,t_1)f(t_1)dt_1dt+......\\ &= y(x)-f(x)\end{aligned}$$

https://doi.org/ 10.2478/9788366675575-002

Thus, we obtain the following.

Theorem-2.1

If

(a) $y(x)=f(x)+\lambda\int_a^b K(x,t)y(t)dt$ (a,b constants)

(b) $K(x,t)$ is real and continuous in a rectangle R for which $a\le x\le b, a\le t\le b, |K(x,t)|\le M$ in $R, K(x,t)\ne 0$.

(c) $f(x)\ne 0$ is real and continuous in the interval $I: a\le x\le b$.

(d) λ is constant and $|\lambda|<\dfrac{1}{M(b-a)}$.

Then the equation (2.1.1) has one and only one continuous solution in I and this solution is given by the absolutely and uniformly convergent series (2.1.3).

The equation

$$y(x)=f(x)+\int_a^b K(x,t)y(t)dt \tag{2.1.4}$$

is a particular case of the equation (2.1.1), for which $\lambda=1$.

The discussion just made holds without change after putting $\lambda=1$.

https://doi.org/ 10.2478/9788366675575-002

Equation (2.1.1) and (2.1.4) may have a continuous solution, even though the hypothesis (d), i.e., $|\lambda| M(b-a) < 1$ is not fulfilled.

The following example shows the truth of this statement:

$$y(x) = \frac{x}{2} - \frac{1}{3} + \int_0^1 (x+t)\, y(t)\, dt$$

Which has the continuous solution $y(x) = x$,

while $|\lambda| M(b-a) = 2$ not less than 1 .

2.2. Volterra Integral Equation:

The equation

$$y(x) = f(x) + \lambda \int_a^x K(x,t)\, y(t)\, dt \qquad (2.2.1)$$

is known as Volterra's integral equation.

Let us substitute successively for $u(t)$ its value as given by (2.2.1).

We find

$$y(x)=f(x)+\lambda\int_a^x K(x,t)f(t)dt$$
$$+\lambda^2\int_a^x K(x,t)\int_a^t K(t,t_1)f(t_1)dt_1dt+.... \tag{2.2.2}$$
$$+\lambda^n\int_a^x K(x,t)\int_a^t K(t,t_1).......$$
$$\int_a^{t_{n-2}} K(t_{n-2},t_{n-1})f(t_{n-1})dt_{n-1}......dt_1dt+R_{n+1}(x)$$

Where

$$R_{n+1}(x)=\lambda^{n+1}\int_a^x K(x,t)\int_a^t K(t,t_1).....$$
$$...\int_a^t K(t_{n-1},t_n)u(t_n)dt_n...dt_1dt$$

We consider the infinite series

$$y(x)=f(x)+\lambda\int_a^x K(x,t)f(t)dt$$
$$+\lambda^2\int_a^x K(x,t)\int_a^t K(t,t_1)f(t_1)dt_1dt+...... \tag{2.2.3}$$

The general term $V_n(x)$ of this series may be written

$$V_n(x)=\lambda^n\int_a^x K(x,t)\int_a^t K(t,t_1).....$$
$$\int_a^{t_{n-2}} K(t_{n-2},t_{n-1})f(t_{n-1})dt_{n-1}...dt_1dt$$

https://doi.org/ 10.2478/9788366675575-002

Then,

Since $|K(x,t)| \le M$ in R and $|f(t)| \le N$ in I,

We have

$$|V_n(x)| \le |\lambda^n| NM^n \frac{(x-a)^n}{n!}$$

$$\le |\lambda^n| N \frac{[M(b-a)]^n}{n!}, (a \le x \le b)$$

The series for which the positive constant

$|\lambda^n| N \frac{[M(b-a)]^n}{n!}$ is the general expression for the nth term

is convergent for all values of $\lambda, N, M, (b-a)$.

Hence the series (2.2.3) is absolutely and uniformly convergent.

If (2.2.1) has a continuous solution, then it must be expressed by (2.2.3).

If $y(x)$ is continuous in I, its absolute value has a maximum value Y, then

$$|R_{n+1}(x)| \le |\lambda^{n+1}| YM^{n+1} \frac{(x-a)^{n+1}}{(n+1)!}$$

$$\le |\lambda^{n+1}| Y \frac{[M(b-a)]^{n+1}}{(n+1)!}$$

https://doi.org/ 10.2478/9788366675575-002

Whence $\lim_{n\to\infty} R_{n+1}(x)=0$.

Thus, we see that the function $y(x)$ satisfying (2.2.2) is the continuous function given by the series (2.2.3).

As before, we can show that the expression for $y(x)$ provided by (2.2.3) satisfies (2.2.1).

Hence, we have the following:

Theorem: -2.2

If

(a) $y(x)=f(x)+\lambda\int_a^x K(x,t)y(t)dt$ (a=constant)

(b) $K(x,t)$ is real and continuous ion the rectangle R, for which $a\le x\le b$, $a\le t\le b$, $|K(x,t)|\le M$ in R, $K(x,t)\ne 0$.

(c) $f(x)\ne 0$ is real and continuous in $I: a\le x\le b$.

(d) λ is constant.

Then the equation (2.2.1) has one, and only one continuous solution $y(x)$ in I and this solution is given by the absolutely and uniformly convergent series (2.2.3).

https://doi.org/ 10.2478/9788366675575-002

The result of this section holds without any change for the equation $y(x)=f(x)+\lambda\int_a^x K(x,t)y(t)dt$, by putting $\lambda=1$.

2.3. Successive approximation:

We would like to point out that the method of solution by successive approximation differs from successive substitutions.

Under the method of successive approximation, we select any real function $y_0(x)$ continuous in *I* . Substitute in the right-hand member of

$$y(x)=f(x)+\lambda\int_a^b K(x,t)y(t)dt \tag{2.3.1}$$

In place of $y(t)$, put the function $y_0(t)$,

It finds $y_1(x)=f(x)+\lambda\int_a^b K(x,t)y_0(t)dt$

The function $y_1(x)$ so determined is real and continuous in *I* .

https://doi.org/ 10.2478/9788366675575-002

Continuous in like manner by replacing y_0 by y_1 and so on. We obtain a series of functions $y_0(x), y_1(x), \ldots\ldots, y_n(x), \ldots$

which satisfy the equations,

$$y_2(x) = f(x) + \lambda \int_a^b K(x,t) y_1(t) dt$$

… … …

$$y_{n-1}(x) = f(x) + \lambda \int_a^b K(x,t) y_{n-2}(t) dt \tag{2.3.2}$$

$$y_n(x) = f(x) + \lambda \int_a^b K(x,t) y_{n-1}(t) dt$$

… … …

From these equations, we find

$$\begin{aligned} y_n(x) = f(x) &+ \lambda \int_a^b K(x,t) f(t) dt \\ &+ \lambda^2 \int_a^b K(x,t) \int_a^b K(t,t_1) f(t_1) dt_1 dt + \ldots. \\ &+ \lambda^{n-1} \int_a^b K(x,t) \int_a^b K(t,t_1) \ldots\ldots. \\ &\int_a^b K(t_{n-3},t_{n-2}) f(t_{n-2}) dt_{n-2} \ldots\ldots dt_1 dt + R_n(x) \end{aligned} \tag{2.3.3}$$

Where

https://doi.org/ 10.2478/9788366675575-002

$$R_n(x)=\lambda^n\int_a^b K(x,t)\int_a^b K(t,t_1).....$$

$$\int_a^b K(t_{n-2},t_{n-1})y_0(t_{n-1})dt_{n-1}...dt_1dt$$

If $y_0(x)$ is real and continuous in I and so has a maximum value Y in I.

Then it is easy to see that

$$|R_n|\le|\lambda^n|YM^n(b-a)^n$$

If then $|\lambda|M(b-a)<1$, we have

$$\lim_{n\to\infty}R_n(x)=0$$

Thus as n-increases, the series of functions $y_n(x)$ approach a limit function given by the series in the right member of (2.3.3). We identify this series with the right member of (2.2.3), Thus

$$\lim_{n\to\infty}y_n(x)\equiv y(x)$$

By this process, at each step, a new function $y_n(x)$ appears dependent upon the choice of $y_0(x)$. We notice, however, that the Limit is independent of the choice of $y_0(x)$

.We can now make an independent proof of the uniqueness of the solution.

Suppose that there is another solution $z(x)$.

Choose $y_0(x) \equiv z(x)$.

It is then clear that each $y_n(x)$ will be identical with $z(x)$ and hence the limit will be $z(x)$. But we have just seen that the limit is independent of the choice of $y_0(x)$. Therefore $z(x) \equiv y(x)$.

A similar discussion can be carried out for the Volterra equation.

2.4. Iterated Functions:

Let us Place

$$K_1(x,t) = K(x,t)$$
$$K_i(x,t) = \int_a^b K(x,s) K_{i-1}(s,t) ds \tag{2.4.1}$$

The function $K_1, K_2, ..., K_n, ..$ so formed are called *Iterated functions*.

By successive approximations of equation (2.4.1), It is evident that

$$K_i(x,t)=\int_a^b......\int_a^b K(x,s_1)K(s_1,s_2).....K(s_{i-1},t)\,ds_{i-1}.....ds_1 \quad (2.4.2)$$

From (24.2), $K_n(x,t)$ is an $(n-1)$ fold integral and $K_p(s,t)$ is a $(p-1)$ fold integral.

Whence we see that

$$\int_a^b K_n(x,s)K_p(s,t)\,ds.$$

is an $(n+p-1)$ fold integral,

which by some changes in the order of integration, is identical to $K_{n+p}(x,t)$.

Hence

$$K_{n+p}(x,t)=\int_a^b K_n(x,s)K_p(s,t)\,ds\ . \quad (2.4.3)$$

2.5. Reciprocal functions:

$$-k(x,t)=K_1(x,t)+K_2(x,t)+.......+K_n(x,t)+.... \quad (2.5.1)$$

It is easy to show that, when $K(x,t)$ is real and continuous in R, The infinite series for $k(x,t)$ is absolutely and uniformly convergent, If $M(b-a)<1$.

https://doi.org/ 10.2478/9788366675575-002

Consequently, $k(x,t)$ is real and continuous in R .

On account of the first of equations (2.4.1) and equation (2.4.3), we have

$$
\begin{aligned}
&-k(x,t)-K(x,t) \\
&= K_2(x,t)+K_3(x,t)+.......+K_n(x,t)+.... \\
&= \int_a^b K_1(x,s)K_1(s,t)\,ds+......+\int_a^b K_1(x,s)K_{n-1}(s,t)\,ds+...... \\
&= \int_a^b K_1(x,s)K_1(s,t)\,ds+......+\int_a^b K_{n-1}(x,s)K_1(s,t)\,ds+......
\end{aligned}
$$

This equation can be written as

$$
\begin{aligned}
-k(x,t)-K(x,t) &= \int_a^b K_1(x,s)\left[K_1(s,t)+.....+K_{n-1}(s,t)+....\right]ds \\
&= \int_a^b \left[K_1(x,s)+.....+K_{n-1}(x,s)+....\right]K_1(s,t)\,ds
\end{aligned}
$$

Upon using equation (2.5.1), we obtain the following characteristics formula:

$$
\begin{aligned}
k(x,t)+K(x,t) &= \int_a^b K(x,s)k(s,t)\,ds \\
&= \int_a^b k(x,s)K(s,t)\,ds
\end{aligned}
\tag{2.5.2}
$$

https://doi.org/ 10.2478/9788366675575-002

Two functions $K(x,t)$ and $k(x,t)$ are *reciprocal* if they are both real and continuous in R and if they satisfy the condition (2.5.2).

A function $k(x,t)$ reciprocal to $K(x,t)$ will exist the series in (2.5.1) converges uniformly.

But we have seen that this series converges uniformly, when $M(b-a)<1$, where M is the maximum of $|K(x,t)|$ in R.

Thus, we have the following:

Theorem: -2.3

If $K(x,t)$ is real and continuous in R , then there exists a reciprocal function $k(x,t)$ given by (2.5.1) provided that $M(b-a)<1$, where M is the maximum of $|K(x,t)|$ in R.

2.6. Volterra Solution of Fredholm's equation:

Volterra has shown how to find a solution of

$$y(x)=f(x)+\int_a^b K(x,t)y(t)dt \tag{2.6.1}$$

Whence the reciprocal function $k(x,t)$ of $K(x,t)$ is known. If (2.6.1) has a continuous solution $y(x)$, then

$$y(t)=f(t)+\int_a^b K(t,t_1)y(t_1)dt_1$$

Multiplying by $k(x,t)$ and integrating, we find

$$\begin{aligned}\int_a^b k(x,t)y(t)dt&=\int_a^b k(x,t)f(t)dt+\int_a^b\int_a^b k(x,t)K(t,t_1)y(t_1)dt_1dt\\&=\int_a^b k(x,t)f(t)dt+\int_a^b\left[K(x,t)+k(x,t_1)\right]y(t_1)dt_1\end{aligned}$$

Which reduces to

$$0=\int_a^b k(x,t)f(t)dt+\int_a^b K(x,t_1)y(t_1)dt_1 \qquad (2.6.2)$$

But from (2.6.1), we have

$$\int_a^b K(x,t_1)y(t_1)dt_1=y(x)-f(x)$$

Therefore (2.6.2) may be written as

$$y(x)=f(x)-\int_a^b k(x,t)f(t)dt \qquad (2.6.3)$$

If (2.6.1) has a continuous solution, then it is given by this above formula, and it is unique.

https://doi.org/ 10.2478/9788366675575-002

The expression for $y(x)$ is given by (2.6.3) can be written in the form

$$f(x) = y(x) + \int_a^b k(x,t) f(t)\,dt$$

This is an integral equation for the determination of $f(x)$.

The function reciprocal to $k(x,t)$ is $K(x,t)$, By what we have just proved.

This equation has a continuous unique solution and is given by $f(x) = y(x) - \int_a^b K(x,t) f(t)\,dt$

Thus, we see that (2.6.1) is satisfied by the value of $y(x)$ given by (2.6.3).

Hence, we have the following.

Theorem: -2.4

If

(a) $K(x,t)$ is real and continuous in R and $K(x,t) \neq 0$.

(b) $f(x)$ is real and continuous in I, $f(x) \neq 0$.

(c) A function $k(x,t)$ *reciprocal* to $K(x,t)$ exists.

Then the equation (2.6.1) has one and only one continuous solution in R, and this solution is given by (2.6.3).

The same reasoning applied to (2.5.2), considered an integral equation for determining $k(x,t)$, shows that it is unique if a continuous reciprocal function exists.

2.7. Discontinuous Solution:

We have shown the existence under the proper assumption of a unique continuous solution for a linear integral equation of the second kind.

This integral may also have, in addition, a discontinuous solution.

To show this, we exhibit the particular equation

$$y(x)=\int_0^x t^{x-1}y(t)\,dt$$

Which has one and only one continuous solution, namely $y(x)\equiv 0$.

We can show, by direct substitution, that this equation also has an infinite number of discontinuous solutions given by $y(x)=Cx^{x-1}$.

Where C is an arbitrary constant and not zero.

https://doi.org/ 10.2478/9788366675575-002

2.8. Fredholm Equations with separable kernels

The kernel of an integral equation is said to be ***separable or degenerate***, if it can be expressed as

$$K(x,t)=\sum_{i=1}^{n} f_i(x)g_i(t). \qquad (2.8.1)$$

Where n is an integer.

Example-1

$\sin(x+t)$ is separable since $\sin(x+t)=\sin x\cos t+\cos x\sin t$. But $\sin xt$ is not separable.

2.8.1. Fredholm Equations

Let us consider the Fredholm integral equation with the separable kernel as

$$y(x)=f(x)+\lambda\int_a^b\left[\sum_{i=1}^{n} f_i(x)g_i(t)\right]y(t)dt$$

$$\Rightarrow y(x)=f(x)+\lambda\sum_{i=1}^{n} c_i f_i(x) \qquad (2.8.2)$$

Where $c_i=\int_a^b g_i(t)y(t)dt \qquad (2.8.3)$

The constants c_i can be obtained by substituting (2.8.2) in (2.8.3).

Example-2

Solve the Fredholm integral equation

$$y(x) = \lambda \int_0^1 (1-3xt)\, y(t)\, dt + x.$$

Solution:

Here $y(x) = x + \lambda(c_1 - 3c_2 x)$

Where $c_1 = \int_0^1 y(t)\,dt$, $c_2 = \int_0^1 t y(t)\,dt$

Now, substituting for $y(t)$, we get

$$2(1-\lambda)c_1 + 3\lambda c_2 = 1$$
$$-3\lambda c_1 + 6(1+\lambda)c_2 = 2$$

This system has a unique solution

$$c_1 = \frac{2}{4-\lambda^2}, c_2 = \frac{4-\lambda}{3(4-\lambda^2)}$$

Provided $\lambda \neq \pm 2$.

If $\lambda = \pm 2$, then the system has no solution.

Hence the desired solution is

https://doi.org/ 10.2478/9788366675575-002

$$y(x)=x+\frac{\lambda}{4-\lambda^2}\left[2-(4-\lambda)x\right].$$

Example-3

Solve the Fredholm integral equation

$$y(x)=1+\int_0^1 xty(t)dt$$

Solution:

Here $y(x)=1+cx$,

Where $c=\int_0^1 ty(t)dt=\int_0^1\left(t+ct^2\right)dt=\frac{1}{2}+\frac{c}{3}$

$$\Rightarrow c=\frac{3}{4}$$

$$\Rightarrow y(x)=1+\frac{3}{4}x$$

Example-4

Solve the Fredholm integral equation

$$y(x)=x+\int_0^1 (x+t)y(t)dt.$$

Solution:

Here $y(x)=(1+c_1)x+c_2$

https://doi.org/ 10.2478/9788366675575-002

Where $c_1 = \int_0^1 y(t)\,dt, c_2 = \int_0^1 ty(t)\,dt$.

$$\Rightarrow c_1 = \frac{1+c_1}{2} + c_2, c_2 = \frac{c_2}{2} + \frac{1+c_1}{3}$$

Solving $c_1 = -7, c_2 = -4$

Hence the general solution is

$$y(x) = -(6x+4)$$

Example-5

Solve the Fredholm integral equation

$$y(x) = 1 + \int_0^1 (x-t)\,y(t)\,dt$$

Solution:

Here $y(x) = 1 - c_2 + c_1 x$

Where

$$c_1 = \int_0^1 y(t)\,dt = 1 - c_2 + \frac{c_1}{2}$$

$$c_2 = \frac{1-c_2}{2} + \frac{c_1}{3}$$

Solving, $c_1 = \frac{12}{13}, c_2 = \frac{7}{13}$

https://doi.org/ 10.2478/9788366675575-002

$$\therefore y(x)=\frac{6}{13}(1+2x)$$

Example-6

Solve the Fredholm integral equation

$$y(x)=1+2\int_{1}^{2}\frac{x}{t}y(t)\,dt$$

Solution:

Here $y(x)=1+2cx$, Where

$$c=\int_{0}^{1}\frac{y(t)}{t}dt=\int_{0}^{1}\left(2c+\frac{1}{t}\right)dt=2c+\ln 2$$

$$\Rightarrow c=-\ln 2$$

Hence $y(x)=1-x\ln 4$.

Example-7

Solve the Fredholm integral equation

$$y(x)=\tan x+\int_{-1}^{1}e^{\sin^{-1}x}y(t)\,dt$$

Solution:

Here $y(x)=\tan x+ce^{\sin^{-1}x}$

Where

$$c=\int_{-1}^{1} y(t)\,dt$$

$$=\int_{-1}^{1} \tan t + ce^{\sin^{-1} t}\,dt$$

$$=c\int_{-1}^{1} e^{\sin^{-1} t}\,dt$$

$$=c\int_{-\frac{\pi}{2}}^{\frac{\pi}{2}} e^{\theta}\cos\theta d\theta$$

$$\therefore c=0$$

Hence $y(x)=\tan x$.

Example-8

Solve the Fredholm integral equation

$$y(x)=\cot x+\lambda\int_{-\frac{\pi}{4}}^{\frac{\pi}{4}} \tan t\, y(t)\,dt.$$

Solution:

Here $y(x)=\cot x+\lambda c$.

Where $c=\int_{-\frac{\pi}{4}}^{\frac{\pi}{4}} \tan t(\cot t+\lambda c)\,dt=\frac{\pi}{2}$.

https://doi.org/ 10.2478/9788366675575-002

$$\therefore y(x)=\cot x+\frac{\lambda\pi}{2}$$

Example-9

Solve the Fredholm integral equation

$$y(x)=\sin x+\lambda\int_0^{\frac{\pi}{2}}\sin x\cos t\, y(t)\,dt\,.$$

Solution:

Here $y(x)=(1+\lambda c)\sin x$

Where

$$c=\int_0^{\frac{\pi}{2}}(1+\lambda c)\sin t\cos t\,dt=\frac{1+\lambda c}{2}$$

$$\therefore c=\frac{1}{2-\lambda}\quad,\ \lambda\neq 2$$

Hence the general solution is $y(x)=\dfrac{2\sin x}{2-\lambda}$.

Example-10

Solve the Fredholm integral equation

$$y(x)=1+\lambda\int_0^1\cos(q\ln t)\,y(t)\,dt\,.$$

Solution:

Here $y(x)=1+\lambda c$

Where

$$c=(1+\lambda c)\int_{0}^{1}\cos(q\ln t)\,dt$$

$$=\frac{1+\lambda c}{1+q^2}$$

$$\therefore c=\frac{1}{1+q^2-\lambda}$$

Hence the general solution is $y(x)=\dfrac{1+q^2}{1+q^2-\lambda}$.

Example-11

Solve the Fredholm integral equation

$$y(x)=x+\lambda\int_{-\pi}^{\pi}\left(x\cos t+t^2\sin x+\cos x\sin t\right)y(t)\,dt$$

Solution:

Here $y(x)=x+c_1\lambda x+c_2\lambda\sin x+c_3\lambda\cos x$

Where

$$c_1=\int_{-\pi}^{\pi}\left(t+c_1\lambda t+c_2\lambda\,\text{sint}+c_3\lambda\,\text{cost}\right)\cos t\,dt$$

$$c_2=\int_{-\pi}^{\pi}\left(t+c_1\lambda t+c_2\lambda\,\text{sint}+c_3\lambda\,\text{cost}\right)t^2\,dt$$

$$c_3 = \int_{-\pi}^{\pi} \left(t + c_1\lambda t + c_2\lambda \,\text{sint} + c_3\lambda \,\text{cost}\right)\sin t dt$$

Evaluating the integrals, we find

$$c_1 - \lambda\pi c_3 = 0$$
$$c_2 + 4\lambda\pi c_3 = 0$$
$$-2\lambda\pi c_1 - \lambda\pi c_2 + c_3 = 2\pi$$

Solving, $c_1 = \dfrac{2\pi^2\lambda}{1+2\pi^2\lambda^2}, c_2 = \dfrac{8\pi^2\lambda}{1+2\pi^2\lambda^2}, \ c_3 = \dfrac{2\pi}{1+2\pi^2\lambda^2}$

$$\therefore y(x) = x + \frac{2\pi\lambda}{1+2\pi^2\lambda^2}\left(\lambda\pi x - 4\lambda\pi \sin x + \cos x\right)$$

2.9. Resolvent Kernel:

Consider the Fredholm integral equation

$$y(x) = f(x) + \lambda\int_a^b K(x,t)\,y(t)\,dt \qquad (2.9.1)$$

We assume that the infinite series give the solution of equation (2.9.1)

$$y(x) = f(x) + \sum_{n=1}^{\infty} z_n(x)\lambda^n \qquad (2.9.2)$$

Where,

https://doi.org/ 10.2478/9788366675575-002

$$z_1(x) = \int_a^b K(x,t) f(t)\,dt$$

$$z_2(x) = \int_a^b K(x,t) z_1(t)\,dt$$

$$= \int_a^b K_2(x,t) f(t)\,dt$$

In general,

$$z_n(x) = \int_a^b K_n(x,t) f(t)\,dt$$

The Iterated kernels $K_1(x,t), K_2(x,t), ..., K_n(x,t)$ are given by

$$K_1(x,t) = K(x,t)$$

$$K_2(x,t) = \int_a^b K(x,\mathrm{s}) K_1(\mathrm{s},t)\,ds$$

$$K_3(x,t) = \int_a^b K(x,\mathrm{s}) K_2(\mathrm{s},t)\,ds$$

$$\vdots$$

$$K_n(x,t) = \int_a^b K(x,\mathrm{s}) K_{n-1}(\mathrm{s},t)\,ds$$

Using these iterated kernels, we construct the Neumann series

$$R(x,t;\lambda) = \sum_{n=1}^{\infty} K_n(x,t) \lambda^{n-1} \tag{2.9.3}$$

https://doi.org/ 10.2478/9788366675575-002

The function $R(x,t;\lambda)$ is called the ***Resolvent kernel***. The solution of the Fredholm integral equation (2.9.1) is given by

$$y(x)=f(x)+\lambda\int_a^b R(x,t;\lambda)f(t)dt . \square$$

Example-12

Find the resolvent kernel for $K(x,t)=xe^t,\ a=-1,b=1$.

Solution:

$$K(x,t)=xe^t$$

$$K_2(x,t)=\int_{-1}^{1} xe^s.se^t ds = xe^t.\frac{2}{e}$$

$$K_3(x,t)=xe^t.\left(\frac{2}{e}\right)^2$$

$$\vdots$$

$$K_n(x,t)=xe^t.\left(\frac{2}{e}\right)^{n-1}$$

$$\therefore R(x,t;\lambda)=K_1(x,t)+K_2(x,t)\lambda+.........$$

$$=xe^t\left[1+\frac{2\lambda}{e}+\frac{4\lambda^2}{e^2}+.......\right]$$

$$=\frac{xe^{t+1}}{e-2\lambda},\quad |\lambda|<\frac{e}{2}$$

https://doi.org/ 10.2478/9788366675575-002

Example-13

Find the resolvent kernel for

$K(x,t)=xt,\quad a=-1,b=1$.

Solution:

$$K(x,t)=xt$$

$$K_2(x,t)=\int_{-1}^{1} xs.st\,ds=\frac{2}{3}xt$$

$$K_3(x,t)=\left(\frac{2}{3}\right)^2 xt$$

$$\vdots$$

$$\therefore R(x,t;\lambda)=K_1(x,t)+K_2(x,t)\lambda+.........$$

$$=xt\left[1+\frac{2\lambda}{3}+\left(\frac{2\lambda}{3}\right)^2+...\right]$$

$$=\frac{3xt}{3-2\lambda},|\lambda|<\frac{3}{2}$$

Example-14

Find the resolvent kernel for

$K(x,t)=xt,\quad a=0,b=1$.

Solution:

https://doi.org/ 10.2478/9788366675575-002

$$K(x,t) = xt$$

$$K_2(x,t) = \int_0^1 xs.st\, ds = \frac{1}{3} xt$$

$$K_3(x,t) = \frac{xt}{3^2}$$

$$\vdots$$

$$\begin{aligned} R(x,t;\lambda) &= K_1(x,t) + K_2(x,t)\lambda + \dots\dots\dots \\ &= xt\left[1 + \frac{\lambda}{3} + \left(\frac{\lambda}{3}\right)^2 + \dots\right] \\ &= \frac{3xt}{3-\lambda},\ |\lambda| < 3 \end{aligned}$$

Example-15

Find the resolvent kernel for

$K(x,t) = e^{x+t},\ a = 0, b = 1.$

Solution:

$$K(x,t) = e^{x+t}$$

$$K_2(x,t) = \int_0^1 e^{x+s}.e^{s+t}\, ds = e^{x+t}.\left(\frac{e^2-1}{2}\right)$$

$$K_3(x,t) = e^{x+t}.\left(\frac{e^2-1}{2}\right)^2$$

$$\vdots$$

https://doi.org/ 10.2478/9788366675575-002

$$R(x,t;\lambda)=K_1(x,t)+K_2(x,t)\lambda+.........$$

$$=e^{xt}\left[1+\frac{\lambda}{2}\left(e^2-1\right)+\frac{\lambda^2}{2^2}\left(e^2-1\right)^2+...\right]$$

$$=\frac{2e^{xt}}{2-\lambda\left(e^2-1\right)},\quad |\lambda|<\frac{2}{e^2-1}$$

Example-16

Find the resolvent kernel for

$$K(x,t)=\sin x\cos t,\ a=0,\ b=\frac{\pi}{2}$$

Solution:

$$K(x,t)=\sin x\cos t$$

$$K_2(x,t)=\int_0^{\frac{\pi}{2}}\sin x\cos s.\sin s\cos t\,ds=\frac{\sin x\cos t}{2}$$

$$K_3(x,t)=\frac{\sin x\cos t}{2^2}$$

$$\vdots$$

$$\therefore R(x,t;\lambda)=K_1(x,t)+K_2(x,t)\lambda+.........$$

$$=\sin x\cos t\left[1+\frac{\lambda}{2}+\frac{\lambda^2}{2^2}+...\right]$$

$$=\frac{2\sin x\cos t}{2-\lambda},\ |\lambda|<2$$

□

https://doi.org/ 10.2478/9788366675575-002

Exercises

1. Solve the linear Fredholm integral equation

$$y(x)=\sec^2(x)+\lambda\int_0^1 y(t)\,dt$$

2. Solve the linear Fredholm integral equation

$$y(x)=\sec^2(x)\tan(x)-\lambda\int_0^1 y(t)\,dt$$

3. Solve the linear Fredholm integral equation

$$y(x)=\cos(x)+\lambda\int_0^\pi xty(t)\,dt$$

4. Solve the linear Fredholm integral equation

$$y(x)=x+\lambda\int_0^1 (x-t)y(t)\,dt$$

5. Solve the non-linear Fredholm integral equation

$$y(x)=\frac{3x}{4}+\int_0^1 xty^2(t)\,dt$$

Solve and find $y(x)$ from the equations.

6. $y(x)=e^x+\lambda\int_0^1 x.t\,y(t)\,dt$

Ans. $y(x)=e^x+\dfrac{3\lambda x}{3-\lambda},\lambda\neq 3$

7. $y(x)=\lambda\int_{0}^{\pi}\sin x\, y(t)\,dt$

Ans. $y(x)=0, if\ \lambda\neq\frac{1}{2}, y(x)=\sin x, if\ \lambda=\frac{1}{2}$

8. $y(x)=\lambda\int_{0}^{\pi}\cos x\, y(t)\,dt$

Ans. $y(x)=0$

9. $y(x)=x+\lambda\int_{0}^{1}(x-t)\,y(t)\,dt$

Ans. $y(x)=\frac{6(\lambda+2)x}{\lambda^2+12}-\frac{4\lambda}{\lambda^2+12}$

10. $y(x)=f(x)+\lambda\int_{0}^{2\pi}\sin x\sin t\, y(t)\,dt$

Ans. $y(x)=f(x)+\frac{\lambda\sin x}{1-\lambda\pi}, \lambda\neq\frac{1}{\pi}$

Construct resolvent kernels for the following kernels.

11. $K(x,t)=\cos(x+t), a=0, b=2\pi$

Ans. $R(x,t;\lambda)=\frac{\cos(x+t)+\pi\lambda\cos(x-t)}{1-\lambda^2\pi^2}$

12. $K(x,t)=1-3xt, a=0, b=1$

Ans. $R(x,t;\lambda)=\dfrac{1+\lambda-\dfrac{3}{2}\lambda(x+t)-3(1-\lambda)xt}{1-\dfrac{\lambda^2}{4}}$

13. $K(x,t)=xt+x^2t^2\ ,a=-1,b=1$

Ans. $R(x,t;\lambda)=\dfrac{3xt}{3-2\lambda}+\dfrac{5x^2t^2}{5-2\lambda}\ ,|\lambda|<\dfrac{3}{2}$

14. $K(x,t)=\sin(x-2t)\ ,a=0,b=2\pi$

Ans. $R(x,t;\lambda)=\sin(x-2t)$

15. $K(x,t)=\sin x\cos t+\cos 2x\sin 2t\ ,a=0,b=2\pi$

Ans. $R(x,t;\lambda)=\sin x\cos t+\cos 2x\sin 2t$

https://doi.org/ 10.2478/9788366675575-003

Chapter-3

Fredholm and Volterra Integral Equations

3.1. Fredholm equation as a limit of a finite system of linear equations:

The solution of the Fredholm equation

$$y(x)=f(x)+\lambda\int_a^b K(x,t)y(t)dt \qquad (3.1.1)$$

has the disadvantage of holding only for restricted values of λ .It is desirable to have, if possible, a solution which holds for all values of λ.such a solution is given by Fredholm in the form

$$y(x)=\frac{\beta_0(x)+\beta_1(x)\lambda+.........}{\alpha_0+\alpha_0\lambda+.......}$$

The numerator and denominator being permanently converging power series in λ .

(a) The system of linear equations replacing the Integral equations:

Before stating explicitly and proving Fredholm's result, we outline the reasoning that led him to his discovery.

Divide the interval (a,b) into n equal parts and call the points of division $t_1,t_2,.....,t_{n-1},..,$

Then

$$t_0 = a, t_1 = a+h, t_2 = a+2h, ..., t_n = a+nh, \tag{3.1.2}$$

$$h = \frac{b-a}{n}$$

Replace the definite integral in (3.1.1) by the sum, corresponding to the points of division (3.1.2), of which it is the Limit.

We obtain the approximate equation

$$y(x) - \lambda h\left[K(x,t_1)y(t_1) + K(x,t_2)y(t_2) + ... \right.$$
$$\left. .... + K(x,t_n)y(t_n)\right] = f(x)$$

Since this equation holds for every value of x, It must be satisfied for $x = t_1, t_2, ..., t_n$.

We thus obtain the following system of n linear equations for the determination of the n unknowns $y(t_1), y(t_2), ..., y(t_n)$:

https://doi.org/ 10.2478/9788366675575-003

$$y(t_1)-\lambda h\left[K(t_1,t_1)y(t_1)+K(t_1,t_2)y(t_2)+........+K(t_1,t_n)y(t_n)\right] = f(t_1)$$

$$y(t_2)-\lambda h\left[K(t_2,t_1)y(t_1)+K(t_2,t_2)y(t_2)+........+K(t_2,t_n)y(t_n)\right] = f(t_2) \quad (3.1.3)$$

....

$$y(t_n)-\lambda h\left[K(t_n,t_1)y(t_1)+K(t_n,t_2)y(t_2)+........+K(t_n,t_n)y(t_n)\right] = f(t_n)$$

This system is solved with respect to $y(t_1),y(t_2),....,y(t_n)$. We can plot $y(t_i)(i=1,2,...,n)$ as ordinates and draw a curve by interpolation, which we may expect to approximate to the real solution. We solve (3.1.3), making use of the abbreviations

$$f(t_i)=f_i,\quad y(t_i)=y_i,\quad K(t_i,t_j)=K_{ij}$$

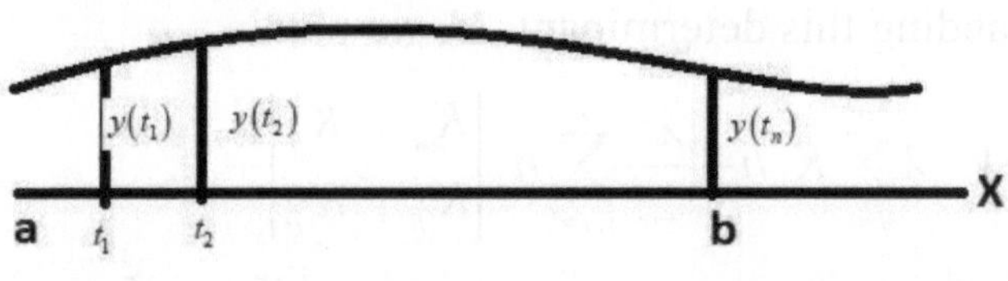

Fig.-1

Denote Δ by the determinant of the coefficient of the u_i in (3.1.3), we have

https://doi.org/ 10.2478/9788366675575-003

$$\Delta = \begin{vmatrix} 1-\lambda hK_{11} & -\lambda hK_{12} & \cdots & -\lambda hK_{1n} \\ -\lambda hK_{21} & 1-\lambda hK_{22} & \cdots & -\lambda hK_{2n} \\ \vdots & \vdots & \vdots & \vdots \\ -\lambda hK_{n1} & -\lambda hK_{n2} & \cdots & 1-\lambda hK_{nn} \end{vmatrix}$$

Denote by $\Delta_{v\mu}$ the first minor of the element in the v row and μ column of Δ.

Then from the Crammers formula, we obtain by solving (3.1.3) with respect to u_k.

$$y_k = \frac{\sum_{i=1}^{n} {}^{i}f_i \Delta_{ik}}{\Delta}, \text{Provided } \Delta \neq 0, k = 1,2,...,n. \qquad (3.1.4)$$

(b) Limit of Δ:

Expanding this determinant Δ, we obtain

$$\Delta = 1 - \lambda \sum_{i=1}^{n} K_{ii} h + \frac{\lambda^2}{2!} \sum_{i,j=1}^{n} h^2 \begin{vmatrix} K_{ii} & K_{ij} \\ K_{ji} & K_{jj} \end{vmatrix} + ...$$

$$... + (-1)^n h^n \frac{\lambda^n}{n!} \begin{vmatrix} K_{11} & K_{12} & \cdots & K_{1n} \\ K_{21} & K_{22} & \cdots & K_{2n} \\ \vdots & \vdots & \vdots & \vdots \\ K_{n1} & K_{n2} & \cdots & K_{nn} \end{vmatrix}$$

https://doi.org/ 10.2478/9788366675575-003

If we are not let n increase indefinitely, we see that each term of this series has a definite limit. So that, at least formally,

$$\lim_{n\to\infty}\Delta = 1-\lambda\int_a^b K(t,t)\,dt$$
$$+\frac{\lambda^2}{2!}\int_a^b\int_a^b\begin{vmatrix} K(t_1,t_1) & K(t_1,t_2) \\ K(t_2,t_1) & K(t_2,t_2)\end{vmatrix}dt_1dt_2 \tag{3.1.5}$$
$$-\frac{\lambda^3}{3!}\int_a^b\int_a^b\int_a^b\begin{vmatrix} K(t_1,t_1) & \vdots & K(t_1,t_3) \\ \cdots & \cdots & \cdots \\ K(t_3,t_1) & \vdots & K(t_3,t_3)\end{vmatrix}dt_1dt_2dt_3$$
$$+\ldots\ldots \equiv D(\lambda)$$

$D(\lambda)$ is called Fredholm's determinant or the determinant of K.

(c) Limit of Δ_{ik}

The expression for $\Delta_{\mu\mu}$ is similar to that for Δ.

$$\Delta_{\mu\mu} = 1-\lambda\sum_{i=1}^{n}{}' K_{ii}h+\frac{\lambda^2}{2!}\sum_{i,j=1}^{n}{}' h^2\begin{vmatrix} K_{ii} & K_{ij} \\ K_{ji} & K_{jj}\end{vmatrix}+\ldots\ldots.$$

Where $'$ means omit $i=\mu$.

Hence

$$\lim_{n\to\infty}\Delta_{\mu\mu} = D(\lambda). \tag{3.1.6}$$

https://doi.org/ 10.2478/9788366675575-003

Again, from rules for expansion of determinants,

$$\Delta_{\nu\mu} = \lambda h\left\{K_{\mu\nu} - \lambda\sum_{i=1}^{n} h\begin{vmatrix} K_{\mu\nu} & K_{\mu i} \\ K_{i\nu} & K_{ii} \end{vmatrix} + \frac{\lambda^2}{2!}\sum_{i,j=1}^{n} h^2 \begin{vmatrix} K_{\mu\nu} & K_{\mu i} & K_{\mu j} \\ K_{i\nu} & K_{ii} & K_{ij} \\ K_{j\nu} & K_{ji} & K_{jj} \end{vmatrix} + \ldots\ldots\right\}$$

Put $hD_{\mu\nu} = \Delta_{\nu\mu}$.

If, as n increases indefinitely, we let (t_μ, t_ν) vary in such a way that $\lim(t_\mu, t_\nu) = (x_1, x_2)$.

We find, at least formally

$$\begin{aligned} \lim_{n\to\infty} D_{\mu\nu} &= \lambda K(x_1,x_2) \\ &- \frac{\lambda^2}{2!}\int_a^b \begin{vmatrix} K(x_1,x_2) & K(x_1,t) \\ K(t,x_2) & K(t,t) \end{vmatrix} dt \\ &+ \frac{\lambda^3}{3!}\int_a^b\int_a^b \begin{vmatrix} K(x_1,x_2) & K(x_1,t_1) & K(x_1,t_2) \\ K(t_1,x_2) & K(t_1,t_1) & K(t_1,t_2) \\ K(t_2,x_2) & K(t_2,t_1) & K(t_2,t_2) \end{vmatrix} dt_1 dt_2 + \ldots\ldots \\ &\equiv D(x_1,x_2;\lambda) \end{aligned} \tag{3.1.7}$$

This expression for $D(x_1, x_2; \lambda)$ is called Fredholm's first minor.

(d) Limit of y_k :

https://doi.org/ 10.2478/9788366675575-003

We can now write (4) in the form

$$y_k = f_k \frac{\Delta_{kk}}{\Delta} + \sum_{i=1}^{n}{}' \frac{f_i \Delta_{ik}}{\Delta} = f_k \frac{\Delta_{kk}}{\Delta} + \sum_{i=1}^{n}{}' \frac{f_i h D_{ki}}{\Delta}$$

Which in the passage to the Limit as $n \to \infty$ becomes

$$y(t_k) = f(t_k) + \frac{1}{D(\lambda)} \int_a^b f(t) D(t_k, t; \lambda) dt$$

But t_k is any point of division. Then we can replace t_k by x and write

$$y(x) = f(x) + \frac{1}{D(\lambda)} \int_a^b f(t) D(x, t; \lambda) dt \qquad (3.1.8)$$

A rigorous mathematical procedure has not obtained this result. However, we are inclined to believe that the above expression $y(x)$ is a solution of (3.1.1).

This belief is later shown to be correct.

(e) Fredholm two fundamental relations:

We will now develop two relations that will be of use to us later in obtaining a solution of the integral equation (3.1.1). We recall the theorem fundamental in the theory of determinants that the sum of the product of the elements of

https://doi.org/ 10.2478/9788366675575-003

any column by the corresponding minors of any other column is zero.

This theorem applied to the determinant Δ gives

$$\left(1-\lambda h K_{jj}\right)\Delta_{jk}-\lambda h K_{kj}\Delta_{kk}-\sum_{i=1}^{n}{}''\lambda h K_{ij}\Delta_{ik}=0$$

Where " means omit $i=j,k$.

Making use of the relation $\Delta_{\nu\mu}=h\mathbf{d}\left(t_{\mu},t_{\nu}\right)$, We find

$$\left(1-\lambda h K_{jj}\right)h\mathbf{d}_{kj}-\lambda h K_{kj}\Delta_{kk}-\sum_{i=1}^{n}\lambda h^{2}K_{ij}\mathbf{d}_{ki}=0$$

Divide through by h, Since $h\neq 0$.

The passage to the Limit as $n\rightarrow\infty$ gives from the last equation by (3.1.6) and (3.1.7)

$$D\left(t_{k},t_{j};\lambda\right)-\lambda K\left(t_{k},t_{j}\right)D(\lambda)-\lambda\int_{a}^{b}K\left(t,t_{j}\right)D\left(t_{k},t;\lambda\right)dt=0\ .$$

This last equation holds for any two points t_j, t_k on the interval (a,b).

Let us put $t_j=x_1$, $t_k=x_2$ and write

$$D\left(x_1,x_2;\lambda\right)-\lambda K\left(x_1,x_2\right)D(\lambda) = \lambda\int_{a}^{b}K\left(t,x_2\right)D\left(x_1,t;\lambda\right)dt \qquad (3.1.9)$$

https://doi.org/ 10.2478/9788366675575-003

This is called *Fredholm's first fundamental relation.*

Now on applying the theorem: The sum of the products of the elements of any row by the corresponding minors of the other row is zero.

This theorem applied to the determinant Δ gives,

$$\left(1-\lambda hK_{jj}\right)\Delta_{kj}-\lambda hK_{jk}\Delta_{kk}-\sum_{i=1}^{n}{}''\lambda hK_{ji}\Delta_{ki}=0$$

Proceeding as before, we find

$$D\left(t_j,t_k;\lambda\right)-\lambda K\left(t_j,t_k\right)D(\lambda)$$
$$-\lambda\int_a^b K\left(t_j,t\right)D\left(t,t_k;\lambda\right)dt=0$$

This equation holds for any two points t_j, t_k on the interval (a,b).

Let us put then $t_j=x_1$, $t_k=x_2$ and write

$$D\left(x_1,x_2;\lambda\right)-\lambda K\left(x_1,x_2\right)D(\lambda)$$
$$-\lambda\int_a^b K\left(x_1,t\right)D\left(t,x_2;\lambda\right)dt=0 \qquad (3.1.10)$$

This is *Fredholm second fundamental relation.*

https://doi.org/ 10.2478/9788366675575-003

3.2. Hadamard's Theorem:

We now proceed to establish the results of the preceding article rigorously. To this end, we need a theorem due to Hadamard.

To establish this theorem, we make use of the following.

Lemma: -3.1

If all of the elements a_{ik} of the determinant

$$A = \begin{vmatrix} a_{11} & a_{12} & \cdots & a_{1n} \\ a_{21} & a_{22} & \cdots & a_{2n} \\ \cdots & \cdots & \vdots & \cdots \\ a_{n1} & a_{n2} & \cdots & a_{nn} \end{vmatrix}$$

are real and satisfy the conditions

$$a_{r1}^2 + a_{r2}^2 + \dots\dots + a_{rn}^2 = 1 , (r = 1,2,3,...,n) \tag{3.2.1}$$

Then $|A| \leq 1$.

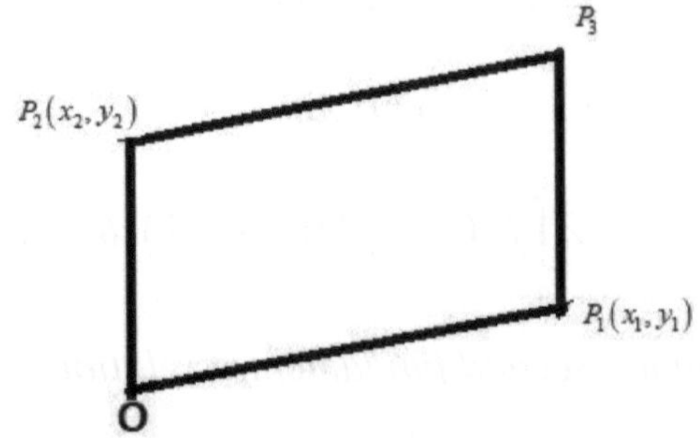

Fig.2

https://doi.org/ 10.2478/9788366675575-003

We give the first two special cases of the lemma, which have a geometric interpretation.

1. $n=2$, The parallelogram $OP_1P_3P_2$ has the vertex O at the origin of a system of rectangular coordinates.

The coordinates of P_1 and P_2 are as indicated in the figure.

The area A of $OP_1P_3P_2$ is given by

$$A=\begin{vmatrix} x_1 & y_1 \\ x_2 & y_2 \end{vmatrix}$$

If $OP_1=OP_2=1$, that is.

If $x_1^2+y_1^2=1$, $x_2^2+y_2^2=1$

Then it is geometrically evident that the greatest area is obtained when the figure is a rectangle, and then the area is 1.

Hence we have generally $|A|\leq 1$.

2. $n=3$, The parallelepiped $OP_1P_2P_3$ has one vertex at the origin of a system of rectangular coordinates. The coordinates of P_1,P_2,P_3 are as indicated in the **figure**-4

The volume V of $OP_1P_2P_3$ is given by

$$V = \begin{vmatrix} x_1 & y_1 & z_1 \\ x_2 & y_2 & z_2 \\ x_3 & y_3 & z_3 \end{vmatrix}.$$

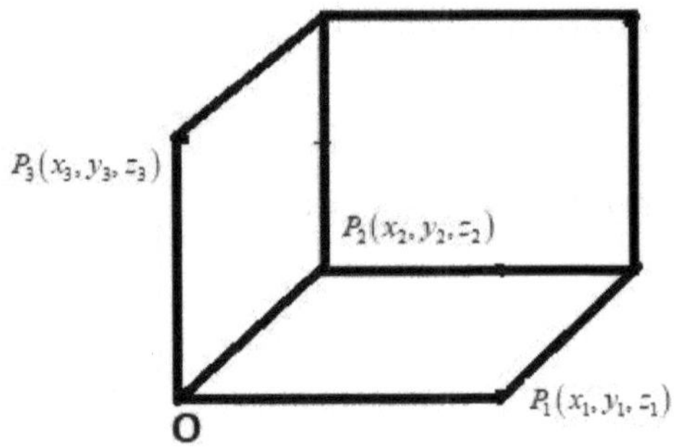

Fig.-3

If $OP_1 = OP_2 = OP_3 = 1$,

That is, If $x_1^2 + y_1^2 + z_1^2 = 1$, $x_2^2 + y_2^2 + z_2^2 = 1, x_3^2 + y_3^2 + z_3^2 = 1$

Then it is evident geometrically that the volume is most significant, when the figure is a rectangular parallelepiped, in which case the volume is 1.

Hence, we have generally $|V| \le 1$.

Proof of the lemma:

$A(a_{11}, a_{12}, ..., a_{1n})$ is a ccontinuous function of its arguments in the region $\Re$ defined by the equation (3.2.1).

https://doi.org/ 10.2478/9788366675575-003

These conditions ensure that $|a_{rs}| \leq 1$ and that the region $\Re$ is bounded and closed. Hence A reaches a *maximum and a minimum* on the region $\Re$. The *maximum and minimum* which we are seeking are the so-called absolute *maximum and minimum*. But if a system of values furnishes the absolute *maximum (minimum)* for A, it also provides a relative *maximum (minimum)*. Hence the ordinary methods of the differential calculus can be used for their determination.

Now, if a function $f(x_1, x_2, ..., x_n)$ of n-variables, connected by h distinct relations.

$\phi_1(x_1, x_2, ..., x_n) = 0$,

$\phi_2(x_1, x_2, ..., x_n) = 0$,

......

$\phi_n(x_1, x_2, ..., x_n) = 0$

has a maximum (minimum), then the n-first partial derivatives of the auxiliary function

$$F = f + \lambda_1\phi_1 + \lambda_2\phi_2 + ... + \lambda_h\phi_h$$

Where $\lambda_1, \lambda_2, ..., \lambda_h$ are constants, must vanish.

For our problem, $f = A, x_r = a_{rs}$

$$\phi_r = \sum_{s=1}^{n} a_{rs}^2 - 1, \ (r = 1, 2, ..., n)$$

and the auxiliary function F becomes

$$F = A + \sum_{s=1}^{n} \frac{\lambda_s}{2}\left(a_{s1}^2 + a_{s2}^2 + \ldots\ldots + a_{sn}^2 - 1\right)$$

For a maximum (minimum) by the theorem just stated, we must have

$$\frac{\partial F}{\partial a_{jk}} = \frac{\partial A}{\partial a_{jk}} + \lambda_j a_{jk} = 0 \text{ , or}$$

$$A_{jk} + \lambda_j a_{jk} = 0, (j, k = 1, 2, \ldots, n) \qquad (3.2.2)$$

Where A_{jk} denotes the minor of a_{jk} in A. Multiply both sides of this equation by a_{jk} and sum with respect to k, for $k = 1, 2, \ldots, n$. We obtain

$$A + \lambda_j = 0,$$

Since $\sum_{k=1}^{n} a_{jk}^2 = 1$

Or $\lambda_j = -A$, $(j = 1, 2, \ldots, n)$

Substitute this value of λ_j in equation (3.2.2)

Then

$$A_{jk} = A a_{jk}, \qquad (j, k = 1, 2, \ldots, n).$$

Whence the determinant

https://doi.org/ 10.2478/9788366675575-003

$$\begin{vmatrix} A_{11} & A_{12} & \cdots & A_{1n} \\ A_{21} & A_{22} & \cdots & A_{2n} \\ \vdots & \vdots & \vdots & \vdots \\ A_{n1} & A_{n1} & \cdots & A_{nn} \end{vmatrix}$$

Adjoint to A, is equal to

$$\begin{vmatrix} Aa_{11} & Aa_{12} & \cdots & Aa_{1n} \\ Aa_{21} & Aa_{22} & \cdots & Aa_{2n} \\ \vdots & \vdots & \vdots & \vdots \\ Aa_{n1} & Aa_{n1} & \cdots & Aa_{nn} \end{vmatrix}$$

The first of these determinants is equal to A^{n-1}.

The second reduces to A^{n+1}.

Hence

$A^{n+1} = A^{n-1}$.

But the *maximum and minimum* of A must satisfy this equation.

Therefore, the maximum of A is +1, and the minimum is -1, and $|A| \leq 1$. □

Hadamard's Theorem:

We are now in a position to prove a more general theorem due to Hadamard:

If the elements b_{ik} of the determinant

$$B=\begin{vmatrix} b_{11} & b_{12} & \cdots & b_{1n} \\ b_{21} & b_{22} & \cdots & b_{2n} \\ \vdots & \vdots & \vdots & \vdots \\ b_{n1} & b_{n1} & \cdots & b_{nn} \end{vmatrix}$$

are real and satisfy the inequality

$|b_{ik}| \le M$, then $|B| \le M^n \sqrt{n^n}$.

Proof:

Let $b_{i1}^2 + b_{i2}^2 + \ldots\ldots + b_{in}^2 = s_i$, $(i = 1, 2, \ldots, n)$

Case-I:

Some one or more of the s_i vanish, say $s_k = 0$.

Then $b_{ki} = 0 \ (i = 1, 2, \ldots\ldots, n)$.

Therefore $B = 0$, and the theorem is proved in this case.

Case-II:

None of the s_i vanish. Then each s_i is positive.

That is $s_1 > 0, s_2 > 0, \ldots\ldots, s_n > 0$.

We now consider the determinant

$$\frac{B}{\sqrt{s_1, s_2, ..., s_n}} = \begin{vmatrix} \frac{b_{11}}{\sqrt{s_1}} & \frac{b_{12}}{\sqrt{s_1}} & \cdots & \frac{b_{1n}}{\sqrt{s_1}} \\ \frac{b_{21}}{\sqrt{s_2}} & \frac{b_{22}}{\sqrt{s_2}} & \cdots & \frac{b_{2n}}{\sqrt{s_2}} \\ \vdots & \vdots & \vdots & \vdots \\ \frac{b_{n1}}{\sqrt{s_n}} & \frac{b_{n1}}{\sqrt{s_n}} & \cdots & \frac{b_{nn}}{\sqrt{s_n}} \end{vmatrix}$$

For which

$$\frac{b_{i1}^2}{s_i} + \frac{b_{i2}^2}{s_i} + ... + \frac{b_{in}^2}{s_i} = 1 \text{ , } (i = 1, 2, ..., n)$$

This determinant satisfies all the condition of our lemma.

Hence $|B| \leq \sqrt{s_1, s_2, ..., s_n}$.

But since $|b_{ik}| \leq M$

We have from

$$s_i = b_{i1}^2 + b_{i2}^2 + ... + b_{in}^2$$

$$s_i \leq nM^2, \ (i = 1, 2, ..., n)$$

Therefore

$|B| \leq M^n \sqrt{n^n}$.

https://doi.org/ 10.2478/9788366675575-003

Convergence Proof:

With the help of Hadamard's theorem, we can now provide the convergence of the series for $D(\lambda)$ and $D(x_1, x_2; \lambda)$.

(a) Convergence of $D(\lambda)$:

$D(\lambda)$ is given by the series

$$D(\lambda) = 1 + \sum_{n=1}^{\infty} (-1)^n \frac{\lambda^n}{n!} A_n , \tag{3.2.3}$$

Where,

$$A_n = \int_a^b \cdots \int_a^b \begin{vmatrix} K(t_1, t_1) & \cdots & K(t_1, t_n) \\ \vdots & \vdots & \vdots \\ K(t_n, t_1) & \cdots & K(t_n, t_n) \end{vmatrix} dt_1 dt_2 \ldots dt_n , \tag{3.2.4}$$

$(n > 0)$

We have assumed that $|K(x,t)| \leq M$ in R. Thus, the determinant in the expression for A_n satisfies all of the condition of Hadamard's theorem, and hence

$$|A_n| \leq \int_a^b \cdots \int_a^b M^n \sqrt{n^n}\, dt_1 \ldots dt_n$$

$$= \sqrt{n^n}\, M^n (b-a)^n$$

Then

$$\left|(-1)^n \frac{\lambda^n}{n!} A_n\right| \leq M^n (b-a)^n \left|\lambda^n\right| \frac{\sqrt{n^n}}{n!}.$$

$$\equiv C_n$$

This identity defines C_n.

We now proceed to show that the series of which C_n is the general term converges. Applying the ratio test, we obtain

$$\frac{C_{n+1}}{C_n} = M(b-a)|\lambda|\sqrt{\left(1+\frac{1}{n}\right)^n}\frac{1}{\sqrt{n+1}}$$

This ratio has for its Limit zero as n becomes infinite. Hence the series of which is the general term converges. Consequently, the series $D(\lambda)$ is absolutely and permanently convergent. □

We state this result in the following:

Theorem: -3.1

The series $D(\lambda)$ is an absolutely and permanently converging power series in λ .

(b) Convergence of $D(x_1,x_2;\lambda)$.

We have

$$D(x_1,x_2;\lambda)=\lambda K(x_1,x_2)+\sum_{n=1}^{\infty}(-1)^n\frac{\lambda^{n+1}}{n!}B_n(x_1,x_2) \qquad (3.2.5)$$

Where

$$B_n(x_1,x_2)$$

$$=\int_a^b\cdots\int_a^b\begin{vmatrix}K(x_1,x_2) & K(x_1,t_1) & \cdots & K(x_1,t_n)\\ K(t_1,x_2) & K(t_1,t_1) & \cdots & K(t_1,t_n)\\ \vdots & \vdots & \vdots & \vdots\\ K(t_n,x_2) & K(t_n,t_1) & \cdots & K(t_n,t_n)\end{vmatrix}dt_1\ldots dt_n \qquad (3.2.6)$$

It is sometimes convenient to write

$$D(x_1,x_2;\lambda)=\sum_{n=0}^{\infty}(-1)^n\frac{\lambda^{n+1}}{n!}B_n(x_1,x_2)$$

Where we consider $B_0(x_1,x_2)=K(x_1,x_2)$ and $0!=1$.

The determinant in the expression for B_n satisfies all of the conditions of Hadamard's theorem and

Hence

$$|B_n|\le\int_a^b\cdots\int_a^b\sqrt{(n+1)^{n+1}}\,dt_1\ldots.dt_n=\sqrt{(n+1)^{n+1}}\,M^{n+1}(b-a)^n$$

Then

https://doi.org/ 10.2478/9788366675575-003

$$\left|(-1)^n \frac{\lambda^{n+1}}{n!} B_n\right| \leq M^{n+1}(b-a)^n |\lambda|^{n+1} \frac{\sqrt{(n+1)^{n+1}}}{n!}$$

$$\equiv E_n$$

The identity defines E_n.

We now proceed to show that the series of which E_n is the general term converges.

Applying the ratio test,

$$\frac{E_n}{E_{n-1}} = M(b-a)|\lambda| \sqrt{\left(1+\frac{1}{n}\right)^n} \cdot \frac{1}{\sqrt{n+1}} \cdot \frac{n+1}{n}$$

The ratio has for its Limit zero as n becomes infinite.

Hence the series of which E_n is the general term converges.

Consequently, the series $D(x_1, x_2; \lambda)$ is absolutely and permanently convergent in λ and uniformly convergent in x_1 and x_2 on R.

Hence the following theorem follows.

https://doi.org/ 10.2478/9788366675575-003

Theorem: -3.2

The series $D(x_1,x_2;\lambda)$ converges absolutely and permanently in λ and uniformly on R: $a \le x_1 \le b, a \le x_2 \le b$.

3.3. Fredholm's Two fundamental Relations

We will now proceed to prove the two relations:

$$D(x_1,x_2;\lambda)-\lambda K(x_1,x_2)D(\lambda)$$

$$=\lambda\int_a^b D(x_1,t;\lambda)K(t,x_2)\,dt \tag{3.3.1}$$

$$=\lambda\int_a^b K(x_1,t)D(t,x_2;\lambda)\,dt \tag{3.3.2}$$

Which were previously heuristically obtained.

(a) Relation between the co-efficient $A_n(x_1,x_2)$ and $B_n(x_1,x_2)$:

Substitute in (3.3.1), in place of $D(x_1,x_2;\lambda)$ and $D(\lambda)$, their series expressions, then both sides of the equality in (3.3.1) become power series in λ.

Hence if (3.3.1) is true, the coefficients of corresponding powers on the two sides must be equal.

https://doi.org/ 10.2478/9788366675575-003

Conversely, if we can show that the coefficients of corresponding powers on the two sides are equal,
Then (3.3.1) will be established.

Making the substitution, we obtain

$$\lambda K(x_1,x_2)+\sum_{n=1}^{\infty}(-1)^n\frac{\lambda^{n+1}}{n!}B_n(x_1,x_2)$$

$$-\lambda K(x_1,x_2)\left[1+\sum_{n=1}^{\infty}(-1)^n A_n\frac{\lambda^n}{n!}\right]$$

$$=\lambda\int_a^b K(t,x_2)\left[\lambda K(x_1,t)+\sum_{n=1}^{\infty}(-1)^n\frac{\lambda^{n+1}}{n!}B_n(x_1,t)\right]dt$$

On the right-hand side, the series in the integrand is uniformly convergent and remains so when multiplied by $K(t,x_2)$.

Therefore, we can integrate by terms and write

$$\lambda^2\int_a^b K(x_1,t)K(t,x_2)\,dt$$

$$+\sum_{n=1}^{\infty}(-1)^n\frac{\lambda^{n+2}}{n!}\int_a^b B_n(x_1,t)K(t,x_2)\,dt$$

If in this second integral we put $n'=n+1$ and then drop the prime, we obtain

$$\sum_{n=1}^{\infty}(-1)^n \frac{\lambda^{n+1}}{n!} B_n(x_1,x_2) - K(x_1,x_2)\sum_{n=1}^{\infty}(-1)^n A_n \frac{\lambda^{n+1}}{n!}$$

$$= \lambda^2 \int_a^b K(x_1,t)K(t,x_2)\,dt$$

$$+\sum_{n=1}^{\infty}(-1)^{n-1}\frac{\lambda^{n+1}}{(n-1)!}\int_a^b B_{n-1}(x_1,t)K(t,x_2)\,dt$$

Compare now the coefficient of corresponding powers of λ on the two sides, we obtain

$$B_n(x_1,x_2) = A_n K(x_1,x_2) - n\int_a^b B_{n-1}(x_1,t)K(t,x_2)\,dt \qquad (3.3.3)$$

If we establish the truth of this equality, then (3.3.1) will be shown to be true.

The relation (3.3.2) treated in precisely the same way leads to the relation

$$B_n(x_1,x_2) = A_n K(x_1,x_2) - n\int_a^b K(x_1,t)B_{n-1}(t,x_2)\,dt \qquad (3.3.4)$$

If we can establish the truth of this equality, then (3.3.2) will be shown to be true.

(b) Proof of (3.3.3) and (3.3.4)

https://doi.org/ 10.2478/9788366675575-003

We prove (3.3.3) and also (3.3.4) by showing that the explicit expression for $B_n(x_1,x_2)$ can by proper expansion and change in notation be written in the form indicated.

The explicit expression for $B_n(x_1,x_2)$ is

$$B_n(x_1,x_2)=\int_a^b \dots \int_a^b \begin{vmatrix} K(x_1,x_2) & K(x_1,t_1) & \cdots & K(x_1,t_n) \\ K(t_1,x_2) & K(t_1,t_1) & \cdots & K(t_1,t_n) \\ \vdots & \vdots & \vdots & \vdots \\ K(t_n,x_2) & K(t_n,t_1) & \cdots & K(t_n,t_n) \end{vmatrix} dt_1 \dots dt_n$$

Develop the determinant which appears in the integrand in terms of the elements of the first column.

We obtain

$$B_n(x_1,x_2)$$

$$=\int_a^b \dots \int_a^b K(x_1,x_2)\times \begin{vmatrix} K(t_1,t_1) & K(t_1,t_2) & \cdots & K(t_1,t_n) \\ K(t_2,t_1) & K(t_2,t_2) & \cdots & K(t_2,t_n) \\ \vdots & \vdots & \vdots & \vdots \\ K(t_n,t_1) & K(t_n,t_2) & \cdots & K(t_n,t_n) \end{vmatrix} dt_1 \dots dt_n$$

$$+\sum_{i=1}^{n}(-1)^i \int_a^b \dots \int_a^b K(t_i,x_2)\times \begin{vmatrix} K(x_1,t_1) & \cdots & K(x_1,t_n) \\ K(t_1,t_1) & \cdots & K(t_1,t_n) \\ \vdots & \vdots & \vdots \\ K(t_{i-1},x_2) & \cdots & K(t_{i-1},t_n) \\ K(t_{i+1},x_2) & \cdots & K(t_{i+1},t_n) \\ \vdots & \cdots & \vdots \\ K(t_n,x_2) & \cdots & K(t_n,t_n) \end{vmatrix} dt_1 \dots dt_n$$

https://doi.org/ 10.2478/9788366675575-003

The first term on the right reduces to $K\left(x_1,x_2\right).A_n$ according to (3.2.4).

In the terms which occur in the summation in place of $t_i,t_{i+1},t_{i+2},....,t_n$

Write $t,t_i,t_{i+1},t_{i+2},....,t_{n-1}$

$$t,t_i,t_{i+1},t_{i+2},....,t_{n-1}\,.$$

Which means simply a change in the notation of the variables of integration in a multiple definite integral.

We obtain

$$\sum_{i=1}^{n}(-1)^i\int_a^b...\int_a^b K\left(t,x_2\right)$$

$$\times\begin{vmatrix} K\left(x_1,t_1\right) & \cdots & K\left(x_1,t_{i-1}\right)K\left(x_1,t\right)K\left(x_1,t_i\right) & \cdots & K\left(x_1,t_{n-1}\right) \\ \cdots & \cdots & \cdots & \cdots & \cdots \\ \cdots & \cdots & \cdots & \cdots & \cdots \\ K\left(x_1,t_{n-1}\right) & \cdots & \cdots & \cdots & K\left(t_{n-1},t_{n-1}\right) \end{vmatrix}$$

$$dt\,dt_1...dt_{n-1}$$

This may be written by bringing into the first column

$$\sum_{i=1}^{n}(-1)^{2i-1}\int_a^b\ldots\int_a^b K(t,x_2)$$

$$\times\begin{vmatrix} K(x_1,t) & K(x_1,t_1) & \cdots & K(x_1,t_{n-1}) \\ K(t_1,t) & K(t_1,t_1) & \cdots & K(t_1,t_{n-1}) \\ \vdots & \vdots & \vdots & \vdots \\ K(t_{n-1},t) & K(t_{n-1},t_1) & \cdots & K(t_{n-1},t_{n-1}) \end{vmatrix}$$

$$dt\, dt_1 \ldots . dt_{n-1}$$

The last expression shows that all of the n–terms of the summation are equal. Furthermore, the integration can be performed in any order.

We then integrate first with respect to $t_1, t_2, t_3, \ldots ., t_{n-1}$ for these integrations, $K(t,x_2)$ may be regarded as a constant factor and may be taken before the $(n-1)$–fold integrals so that we obtain

$$-n\int_a^b K(t,x_2)$$

$$\left\{\int_a^b\ldots\int_a^b \times\begin{vmatrix} K(x_1,t) & K(x_1,t_1) & \cdots & K(x_1,t_{n-1}) \\ K(t_1,t) & K(t_1,t_1) & \cdots & K(t_1,t_{n-1}) \\ \vdots & \vdots & \vdots & \vdots \\ K(t_{n-1},t) & K(t_{n-1},t_1) & \cdots & K(t_{n-1},t_{n-1}) \end{vmatrix} dt_1 \ldots .. dt_{n-1}\right\} dt$$

Which, according to (3.2.6), maybe written

https://doi.org/ 10.2478/9788366675575-003

$$-n\int_a^b B_{n-1}(x_1,t)K(t,x_2)dt$$

Hence the development of $B_n(x_1,x_2)$ gives

$$B_n(x_1,x_2)=A_nK(x_1,x_2)-n\int_a^b B_{n-1}(x_1,t)K(t,x_2)dt$$

Which proves the relation (3.3.3).

Therefore (3.3.1), which is Fredholm's first relation, is true.

Retake the explicit expression $B_n(x_1,x_2)$, but this time, develop the determinant in the integrand in terms of the first-row elements.

Then proceeding as before, we obtain for the development of $B_n(x_1,x_2)$

$$B_n(x_1,x_2)=A_nK(x_1,x_2)-n\int_a^b K(x_1,t)B_{n-1}(t,x_2)dt$$

Which proves the relation (3.3.4).

Therefore (3.3.2), which is Fredholm's second relation, is true.

Thus, we established the following.

https://doi.org/ 10.2478/9788366675575-003

Theorem-3.3

Between Fredholm's determinant $D(\lambda)$ and Fredholm's first minor $D(x_1,x_2;\lambda)$, the following double relation holds:

$$D(x_1,x_2;\lambda)-\lambda K(x_1,x_2)D(\lambda)=\lambda\int_a^b D(x_1,t;\lambda)K(t,x_2)\,dt$$
$$=\lambda\int_a^b K(x_1,t)D(t,x_2;\lambda)\,dt$$

For all values of λ and for all values x_1 and x_2 on R.

3.4. Fredholm's solution of the Integral equation when $D(\lambda)\neq 0$:

Fredholm's two fundamental relations, the equations (3.3.1) and (3.3.2) enables us now to obtain a solution of the integral equation

$$y(x)=f(x)+\lambda\int_a^b K(x,t)y(t)\,dt \tag{3.4.1}$$

We receive a hint as to the method of the procedure from the process of solving the finite system

$$y_i-\lambda h\sum_{j=1}^{n}K_{ij}y_i=f_i,\ (i=1,2,..,n) \tag{3.4.2}$$

To find y_k from this system we first multiply by Δ_{ik} and sum with respect to i, we obtain

$$\sum_{i=1}^{n} y_i\Delta_{ik} - \lambda h\sum_{i=1}^{n}\sum_{j=1}^{n} K_{ij}y_j\Delta_{ik} = \sum_{i=1}^{n} f_i\Delta_{ik}$$

Whence

$$\Delta y_k = \sum_{i=1}^{n} f_i\Delta_{ik} . \tag{3.4.3}$$

Now $\Delta_{ik} = hd_{ki}$ (by definition)

and $\lim_{h\to 0}\sum_{i=1}^{n} h\mathbf{d}_{ki} = \int_a^b D(x,t;\lambda)\,dt$ (by (3.1.6))

Let us now follow the analogy. Write (3.4.1) in the form

$$y(t) = f(t) + \lambda\int_a^b K(t,\xi)\,y(\xi)\,d\xi$$

Multiply both sides of this equation, which we suppose is satisfied by a continuous function y, by $D(x,t;\lambda)$ and then integrate with respect to t from a to b, we obtain

$$\int_a^b D(x,t;\lambda)\,y(t)\,dt = \int_a^b D(x,t;\lambda)\,f(t)\,dt + \lambda\int_a^b\int_a^b D(x,t;\lambda)\,K(t,\xi)\,y(\xi)\,d\xi\,dt \tag{3.4.4}$$

The integrand of the double integral being continuous in x and ξ, we can interchange the order of integration in the double integral and write it

$$\int_a^b y(\xi)\left[\lambda\int_a^b D(x,t;\lambda)K(t,\xi)\,dt\right]d\xi,$$

Which, according to (3.3.1), becomes

$$\int_a^b\left[D(x,\xi;\lambda)-\lambda D(\lambda)K(x,\xi)\right]y(\xi)\,d\xi$$

Hence (3.4.4) may be written

$$\int_a^b D(x,t;\lambda)y(t)\,dt=\int_a^b D(x,t;\lambda)f(t)\,dt$$

$$+\int_a^b D(x,\xi;\lambda)y(\xi)\,d\xi-\lambda D(\lambda)\int_a^b K(x,\xi)y(\xi)\,d\xi$$

Which on account of reduces to

$$0=\int_a^b D(x,t;\lambda)f(t)\,dt-D(\lambda)\left[y(x)-f(x)\right]$$

We now solve for $y(x)$, under the assumption $D(\lambda)\neq 0$.

We obtain

$$y(x)=f(x)+\int_a^b\frac{D(x,t;\lambda)f(t)}{D(\lambda)}\,dt \qquad (3.4.5)$$

Hence, if $y(x)$ is a continuous function of x, that satisfies (3.4.1) and if $D(\lambda)\neq 0$ then it is given by (3.4.5).

It remains for us to show that conversely, the expression $y(x)$ given by (3.4.5) is a solution (3.4.1).

We do this by direct substitution, substitute the value of $y(x)$ (3.4.5) in (3.4.1). we obtain

$$f(x)+\int_a^b \frac{D(x,t;\lambda)f(t)}{D(\lambda)}dt$$

$$=f(x)+\lambda\int_a^b K(x,t)\left\{f(t)+\int_a^b \frac{D(t,\xi;\lambda)f(\xi)}{D(\lambda)}d\xi\right\}dt$$

Break the last term up into two parts and in the double integral change the order of integration.

We obtain

$$\int_a^b \frac{D(x,t;\lambda)f(t)}{D(\lambda)}dt=\lambda\int_a^b K(x,t)f(t)dt$$

$$+\frac{1}{D(\lambda)}\int_a^b f(\xi)\left\{\lambda\int_a^b K(x,t)D(t,\xi;\lambda)dt\right\}d\xi$$

Which, according to (3.3.2), may be written

https://doi.org/ 10.2478/9788366675575-003

$$\int_a^b \frac{D(x,t;\lambda)f(t)}{D(\lambda)}dt = \lambda\int_a^b K(x,t)f(t)dt$$

$$+\frac{1}{D(\lambda)}\int_a^b f(\xi)\left[D(x,\xi;\lambda)-\lambda K(x,\xi)D(\lambda)\right]d\xi$$

But this last equation is seen to be an identity.

Consequently, the expression $y(x)$ given by (3.4.5) satisfies equation (3.4.1).

Thus, we have proved the following theorem, which is called *Fredholm first fundamental theorem.*

Theorem:3.4

If

(a) $D(\lambda)\neq 0$

(b) $K(x,t)$ is continuous in R.

(c) $f(x)$ is continuous in *I* .

Then the equation

$$y(x)=f(x)+\lambda\int_a^b K(x,t)y(t)dt \qquad (3.4.1)$$

has one and only one continuous solution given by

$$y(x)=f(x)+\int_a^b \frac{D(x,t;\lambda)f(t)}{D(\lambda)}dt \qquad (3.4.5)$$

https://doi.org/ 10.2478/9788366675575-003

Where $D(x,t;\lambda)$ and $D(\lambda)$ are absolutely and permanently convergent integral series in λ and $D(x,t;\lambda)$ converges uniformly with respect to x and t on $R: a \le x \le b \quad a \le t \le b$.□

We have at once, for the particular case $f \equiv 0$, the following:

Corollary-3.1

If $D(\lambda) \ne 0$, then the homogenous equation

$$y(x) = \lambda \int_a^b K(x,t)\, y(t)\, dt \tag{3.4.6}$$

has one and only one continuous solution given by $y(x) \equiv 0$.

Let us point out the analogy with the finite system of linear equations

$$y_i - \lambda h \sum_{j=1}^{n} K_{ij} y_j = f_i \quad , \qquad (i = 1, 2, .., n)$$

With determinant Δ.

If $\Delta \ne 0$, then this system has one and only one solution.

If $f_j \equiv 0$, then the only solution is the trivial one

$y_1 = y_2 = ... = y_n \equiv 0$.

But the Limit of Δ was $D(\lambda)$.

https://doi.org/ 10.2478/9788366675575-003

Hence the results of ***theorem-3.4*** and its corollary are exactly what was to be expected from the analogy with the finite system.

3.5. Solution of the Homogenous equation when $D(\lambda)=0$, $D'(\lambda)\neq 0$:

The discussion up to this point has been made under the assumption $D(\lambda)\neq 0$.

Let us now see what happens, when $D(\lambda)=0$, First with respect to the homogenous equation (3.4.6).

Let λ_0 be a value of λ for which

$$D(\lambda_0)=0. \tag{3.5.1}$$

We now consider the solution of the homogenous integral equation (3.4.6) for this particular value of λ:

$$y(x)=\lambda_0\int_a^b K(x,t)\,y(t)\,dt \tag{3.5.2}$$

We obtain a solution of (3.5.2) using Fredholm's second fundamental relation (3.3.2) , which is true for all values of λ and hence for $\lambda=\lambda_0$.

With this value of λ, on account of (3.5.1), equation (3.3.2) becomes

$$D(x_1,x_2;\lambda_0)=\lambda_0\int_a^b K(x_1,t)D(t,x_2;\lambda_0)dt$$

The equality holds for every value of x_2 on the interval (a,b) and therefore for $x_2 = x_{2,0}$. Then

$$D(x_1,x_{2,0};\lambda_0)=\lambda_0\int_a^b K(x,t)D(t,x_{2,0};\lambda_0)dt$$

But this is just the equation (3.5.2) $y(x)$ with replaced by $D(x_1,x_{2,0};\lambda_0)$.

Thus, we see that $y(x)=D(x_1,x_{2,0};\lambda_0)$ is a solution of (3.5.2).

Moreover, this solution is continuous for $D(x_1,x_2;\lambda)$ is uniformly convergent in x and y and its terms are continuous.

But $D(x_1,x_{2,0};\lambda_0)$ maybe identically zero in x_1, either on account of an unfortunate choice of $x_{2,0}$, in which case we could choose some other value for $x_{2,0}$ or because $D(x_1,x_2;\lambda_0)\equiv 0$ in x_1 and x_2, in which case the above

https://doi.org/ 10.2478/9788366675575-003

solution reduces to the trivial one $y \equiv 0$, no matter how we choose $x_{2,0}$.

We have thus proved the following:

Theorem-3.5

If $D(\lambda_0) = 0$ and $D(x_1, x_2; \lambda_0) \neq 0$, Then for a proper choice of y_0, $y(x) = D(x_1, x_{2,0}; \lambda_0)$ is a continuous solution of $y(x) = \lambda_0 \int_a^b K(x,t) y(t) dt$ and $y(x) \neq 0$.

In the theorem just stated, the conditions may replace the condition $D(x_1, x_2; \lambda_0) \neq 0$ by $D'(\lambda_0) \neq 0$.

To show this, we prove the following

$$\int_a^b D(x_1, x_1; \lambda) dx_1 = -\lambda D'(\lambda) \tag{3.5.3}$$

We prove (3.5.3) by making use of the series expression for $D'(\lambda)$ and $D(x_1, x_2; \lambda)$.

We have from (3.2.3)

$$D(\lambda) = 1 - \lambda A_1 + \frac{\lambda^2}{2!} A_2 - \frac{\lambda^3}{3!} A_3 + \ldots.$$

Where the A_n's are given by (3.2.4).

https://doi.org/ 10.2478/9788366675575-003

Then

$$D'(\lambda) = -A_1 + \lambda A_2 - \frac{\lambda^2}{2!} A_3 + ...$$

$$= -\sum_{n=0}^{\infty} (-1)^n \frac{\lambda^n}{n!} A_{n+1}$$

In the expression for A_{n+1} :

$$A_{n+1} = \int_a^b .... \int_a^b \begin{vmatrix} K(t_1,t_1) & K(t_1,t_2) & \cdots & K(t_1,t_{n+1}) \\ K(t_2,t_1) & K(t_2,t_2) & \cdots & K(t_2,t_{n+1}) \\ \vdots & \vdots & \vdots & \vdots \\ K(t_{n+1},t_1) & K(t_{n+1},t_2) & \cdots & K(t_{n+1},t_{n+1}) \end{vmatrix} dt_1....dt_{n+1}$$

In place of $t_1, t_2, t_3, ..., t_n, t_{n+1}$

Put $x_1, t_1, t_2, t_3, ..., t_{n-1}, t_n$

Then

$$A_{n+1} = \int_a^b .... \int_a^b \begin{vmatrix} K(x_1,x_1) & K(x_1,t_1) & \cdots & K(x_1,t_n) \\ K(t_1,x_1) & K(t_1,t_1) & \cdots & K(t_1,t_n) \\ \vdots & \vdots & \vdots & \vdots \\ K(t_n,x_1) & K(t_n,t_1) & \cdots & K(t_n,t_n) \end{vmatrix} dx_1 dt_1....dt_n$$

In this multiple definite integral, we change the order of integration and integrate first with respect to $dt_1 dt_2 ....dt_n$.

https://doi.org/ 10.2478/9788366675575-003

Whence,

A_{n+1}

$$= \int_a^b \left\{ \int_a^b \ldots \int_a^b \begin{vmatrix} K(x_1,x_1) & K(x_1,t_1) & \cdots & K(x_1,t_n) \\ K(t_1,x_1) & K(t_1,t_1) & \cdots & K(t_1,t_n) \\ \vdots & \vdots & \vdots & \vdots \\ K(t_n,x_1) & K(t_n,t_1) & \cdots & K(t_n,t_n) \end{vmatrix} dt_1 \ldots dt_n \right\} dx_1$$

Which on account of (3.2.6) becomes

$$A_{n+1} = \int_a^b B_n(x_1,x_1)\,dx_1 .$$

Therefore

$$D'(\lambda) = -\sum_{n=0}^{\infty} (-1)^n \frac{\lambda^n}{n!} \int_a^b B_n(x_1,x_1)\,dx$$

But $\sum_{n=0}^{\infty} (-1)^n \frac{\lambda^n}{n!} \int_a^b B_n(x_1,x_1)\,dx_1$

is a series uniformly convergent in x_1.

We can then interchange the order of writing the summation and the integration in the expression for $D'(\lambda)$ and write

$$D'(\lambda) = -\sum_{n=0}^{\infty} (-1)^n \frac{\lambda^n}{n!} \int_a^b B_n(x_1,x_1)\,dx_1$$

https://doi.org/ 10.2478/9788366675575-003

Multiply both sides by $-\lambda$ and then by (3.2.5) , we see that, indeed

$$\int_a^b D(x_1,x_1;\lambda)\,dx_1 = -\lambda D'(\lambda) \tag{3.5.4}$$

Let us suppose now that $D(\lambda_0)=0$ and $D'(\lambda_0)\neq 0$, Then certainly $\lambda_0 \neq 0$, since $D(0)=1$.

Hence, if we write (3.5.4) with $\lambda=\lambda_0$, the right-hand side of this equality is different from zero and the left-hand side and, therefore $D(x_1,x_1;\lambda_0)\neq 0$, in x_1 and consequently $D(x_1,x_2;\lambda_0)\neq 0$ in x_1 and x_2 .

Hence indeed, in **theorem-3.5**, the condition $D(x_1,x_2;\lambda_0)\neq 0$ may be replaced by $D'(\lambda_0)\neq 0$.

We remark further that If $y(x)=D(x_1,x_{2,0};\lambda_0)$ is a solution of the homogenous integral equation (3.5.2) then cy (c an arbitrary constant) is also a solution.
Thus, there is an infinitude of solutions that differ only by a constant factor.

We shall later show that there are no other solutions. This is in analogy with the finite system of linear equations

https://doi.org/ 10.2478/9788366675575-003

$$y_i - \lambda h \sum_{j=1}^{n} K_{ij} y_j = 0,\ (i = 1,2,...n)$$

With determinant Δ.

If $\Delta \equiv 0$, while not all of the first minors Δ_{ik} vanish, then these n–equations determine the ratios of them uniquely $y_1, y_2, ..., y_n$ .that is $y_j = c_j y_n$, $(j = 1,2,..,n, c_n = 1)$.

Now $\Delta \equiv 0$ corresponds to $D(\lambda) = 0$, while not all Δ_{ik} vanish corresponds to $D(x_1, x_2; \lambda_0) \neq 0$.□

3.6. Solution of the Homogenous Integral equation when $D(\lambda) \equiv 0$:

It remains to consider the case

$D(\lambda_0) \equiv 0, D(x_1, x_2; \lambda_0) = 0$

This corresponds in the linear system to the case where Δ all its first minors are zero and where it becomes necessary to consider the minors of the higher-order Δ.

Accordingly, we have to consider the integral equation's treatment, limiting these higher minors of Δ as h approaches zero.

In what follow: Let us use with Heywoods-Frechet,

The notation

https://doi.org/ 10.2478/9788366675575-003

$$K\begin{pmatrix} s_1, s_2, \ldots\ldots, s_n \\ t_1, t_2, \ldots\ldots, t_n \end{pmatrix} = \begin{vmatrix} K(s_1,t_1) & K(s_1,t_2) & \cdots & K(s_1,t_n) \\ K(s_2,t_1) & K(s_2,t_2) & \cdots & K(s_2,t_n) \\ \cdots & \cdots & \cdots & \cdots \\ K(s_n,t_1) & K(s_n,t_2) & \cdots & K(s_n,t_n) \end{vmatrix}$$

(a) Definition of the pth minor of $D(\lambda)$:

Let

$$B_n\begin{pmatrix} x_{11}, x_{12}, \ldots\ldots, x_{1p} \\ x_{21}, x_{22}, \ldots\ldots, x_{2p} \end{pmatrix}$$

$$= \int_a^b \ldots \int_a^b K\begin{pmatrix} x_{11}, x_{12}, \ldots\ldots, x_{1p}, t_1, t_2, \ldots\ldots, t_n \\ x_{21}, x_{22}, \ldots\ldots, x_{2p}, t_1, t_2, \ldots\ldots, t_n \end{pmatrix} dt_1 dt_2, \ldots\ldots dt_n \tag{3.6.1}$$

With

$$B_0\begin{pmatrix} x_{11}, x_{12}, \ldots\ldots, x_{1p} \\ x_{21}, x_{22}, \ldots\ldots, x_{2p} \end{pmatrix} = K\begin{pmatrix} x_{11}, x_{12}, \ldots\ldots, x_{1p} \\ x_{21}, x_{22}, \ldots\ldots, x_{2p} \end{pmatrix} \tag{3.6.2}$$

Then, the p-th minor of $D(\lambda)$ is defined by the infinite series

$$D\begin{pmatrix} x_{11}, x_{12}, \ldots\ldots, x_{1p} \\ x_{21}, x_{22}, \ldots\ldots, x_{2p} \end{pmatrix} = \sum_{n=0}^{\infty} (-1)^n \frac{\lambda^{n+p}}{n!} B_n\begin{pmatrix} x_{11}, x_{12}, \ldots\ldots, x_{1p} \\ x_{21}, x_{22}, \ldots\ldots, x_{2p} \end{pmatrix}$$
$$= D_p(x_1, x_2; \lambda)$$

Which for $p = 1$ reduces to $D(x_1, x_2; \lambda)$.

https://doi.org/ 10.2478/9788366675575-003

3.7. Characteristic Constant and Fundamental Functions:

Definition:

If $D(\lambda)$is Fredholm's determinant for the kernel $K(x,t)$and $D(\lambda_0)=0$, then, λ is said to be a *characteristic constant* of the kernel $K(x,t)$.

Further, if $\phi(x)$is continuous and not identically zero on the interval $(a,b)$and

$$\phi(x)=\lambda_0\int_a^b K(x,t)\phi(t)\,dt$$

Then $\phi(x)$is called a *fundamental function* of the kernel $K(x,t)$ belonging to the characteristic constant λ_0.

$\phi_1(x),\phi_2(x),......,\phi_q(x)$ form a complete system of the kernel's $K(x,t)$fundamental function λ_0, If every other solution is expressible linearly in terms of these q solutions.

Thus if $\psi_1,\psi_2,.....,\psi_q$ are any other q solutions. we must have

$$\psi_1=c_{11}\phi_1+c_{12}\phi_2+.....+c_{1q}\phi_q$$

$$\ldots \quad \ldots. \quad \ldots.$$

$$\psi_q=c_{q1}\phi_1+c_{q2}\phi_2+.....+c_{qq}\phi_q$$

https://doi.org/ 10.2478/9788366675575-003

And if

$$\begin{vmatrix} c_{11} & c_{12} & \cdots & c_{1q} \\ c_{21} & c_{22} & \cdots & c_{2q} \\ \cdots & \cdots & \vdots & \cdots \\ c_{q1} & c_{q2} & \cdots & c_{qq} \end{vmatrix} \neq 0$$

$\psi_1, \psi_2, \ldots\ldots, \psi_q$ form again a complete system of fundamental functions belonging to λ_0.

3.8. The Associated Homogenous Integral equations:

Preparatory to the discussion of the non-homogenous integral equation, for $D(\lambda)=0$, we will discuss the homogenous integral equation

$$z(x)=\lambda_0\int_a^b K(t,x)z(t)dt \tag{3.8.1}$$

Which is called the integral equation *associated* with the integral equation

$$y(x)=\lambda_0\int_a^b K(x,t)y(t)dt \tag{3.8.2}$$

Notice that the kernel $\overline{K}(x,t)\equiv K(t,x)$ of the associated equation is derived from the original kernel $K(x,t)$ by interchanging the arguments x and t. There exist essential

https://doi.org/ 10.2478/9788366675575-003

relations between the solution of the two equations (3.8.1) and (3.8.2).

To obtain them, we first compute the Fredholm determinant and Fredholm minors for the kernel $\overline{K}(x,t)$, which we indicate by the corresponding dashed notation.

(a) Fredholm's Determinant for the associated Kernel:

$$D(\lambda)=1+\sum_{n=1}^{\infty}(-1)^n\frac{\lambda^n}{n!}A_n$$

Where

$$A_n=\int_a^b\cdots\int_a^b\begin{vmatrix}K(t_1,t_1) & K(t_1,t_2) & \cdots & K(t_1,t_n)\\ K(t_2,t_1) & K(t_2,t_2) & \cdots & K(t_2,t_n)\\ \vdots & \vdots & \vdots & \vdots\\ K(t_n,t_1) & K(t_n,t_2) & \cdots & K(t_n,t_n)\end{vmatrix}dt_1dt_2....dt_n$$

Then

$$\overline{D}(\lambda)=1+\sum_{n=1}^{\infty}(-1)^n\frac{\lambda^n}{n!}\overline{A_n}$$

Where

https://doi.org/ 10.2478/9788366675575-003

$$\overline{A_n}=\int_a^b\cdots\int_a^b\begin{vmatrix}\overline{K}(t_1,t_1) & \overline{K}(t_1,t_2) & \cdots & \overline{K}(t_1,t_n)\\ \overline{K}(t_2,t_1) & \overline{K}(t_2,t_2) & \cdots & \overline{K}(t_2,t_n)\\ \vdots & \vdots & \vdots & \vdots\\ \overline{K}(t_n,t_1) & \overline{K}(t_n,t_2) & \cdots & \overline{K}(t_n,t_n)\end{vmatrix} dt_1dt_2....dt_n$$

In this expression for $\overline{A_n}$, Put $\overline{K}(x,t)\equiv K(t,x)$

Then

$$\overline{A_n}=\int_a^b\cdots\int_a^b\begin{vmatrix}K(t_1,t_1) & K(t_2,t_1) & \cdots & K(t_n,t_1)\\ K(t_1,t_2) & K(t_2,t_2) & \cdots & K(t_n,t_2)\\ \vdots & \vdots & \vdots & \vdots\\ K(t_1,t_n) & K(t_2,t_n) & \cdots & K(t_n,t_n)\end{vmatrix} dt_1dt_2....dt_n$$

The determinant that appears in the integrand $\overline{A_n}$ is the same as that which appears in the integrand of the expression for A_n, except that the rows and columns are interchanged. Still, this interchange leaves the value of the determinant unaltered.

Therefore $\overline{A_n}=A_n$, and hence

$$\overline{D}(\lambda)\equiv D(\lambda) \tag{3.8.3}$$

Thus, also we conclude that $K(t,x)$ and $K(x,t)$ have the same characteristic constants.

(b) Fredholm's Minors for the Associated Kernel:

We have from (3.6.3),

$$D\begin{pmatrix} x_{11}, x_{12}, \ldots\ldots, x_{1p} & \lambda \\ x_{21}, x_{22}, \ldots\ldots, x_{2p} & \end{pmatrix}$$

$$=\sum_{n=0}^{\infty}(-1)^n \frac{\lambda^{n+p}}{n!} B_n\begin{pmatrix} x_{11}, x_{12}, \ldots\ldots, x_{1p} \\ x_{21}, x_{22}, \ldots\ldots, x_{2p} \end{pmatrix}$$

Where the B_n is given by (3.6.1). Then

$$\overline{B_n}\begin{pmatrix} x_{11}, x_{12}, \ldots\ldots, x_{1p} \\ x_{21}, x_{22}, \ldots\ldots, x_{2p} \end{pmatrix}$$

$$=\int_a^b \ldots\int_a^b \begin{vmatrix} \overline{K}(x_{11},x_{21}) & \cdots & \overline{K}(x_{11},x_{2p})\overline{K}(x_{11},t_1) & \cdots & \overline{K}(x_{11},t_n) \\ \cdots & \cdots & \vdots & \cdots & \cdots \\ \overline{K}(x_{1p},x_{21}) & \cdots & \overline{K}(x_{1p},x_{2p})\overline{K}(x_{1p},t_1) & \cdots & \overline{K}(x_{1p},t_n) \\ \overline{K}(t_1,x_{21}) & \cdots & \overline{K}(t_1,x_{2p})K(t_1,t_1) & \cdots & \overline{K}(t_1,t_n) \\ \cdots & \cdots & \vdots & \cdots & \vdots \\ \overline{K}(t_n,x_{21}) & \cdots & \overline{K}(t_n,x_{2p})\overline{K}(t_n,t_1) & \cdots & \overline{K}(t_n,t_n) \end{vmatrix} dt_1 dt_2 \ldots. dt_n$$

But $\overline{K}(t,x)=K(t,x)$, Then

https://doi.org/ 10.2478/9788366675575-003

$$\overline{B_n}\begin{pmatrix} x_{11}, x_{12}, \ldots\ldots, x_{1p} \\ x_{21}, x_{22}, \ldots\ldots, x_{2p} \end{pmatrix}$$

$$= \int_a^b \ldots \int_a^b \begin{vmatrix} K(x_{21}, x_{11}) & \cdots & K(x_{2p}, x_{11}) K(t_1, x_{11}) & \cdots & K(t_n, x_{11}) \\ \cdots & \cdots & \vdots & \cdots & \cdots \\ K(x_{21}, x_{1p}) & \cdots & K(x_{2p}, x_{1p}) K(t_1, x_{1p}) & \cdots & K(t_n, x_{1p}) \\ K(x_{21}, t_1) & \cdots & K(x_{2p}, t_1) K(t_1, t_1) & \cdots & K(t_n, t_1) \\ \cdots & \cdots & \vdots & \cdots & \vdots \\ K(x_{21}, t_n) & \cdots & K(x_{2p}, t_n) \overline{K}(t_1, t_n) & \cdots & K(t_n, t_n) \end{vmatrix} dt_1 dt_2 \ldots dt_n$$

An interchange of rows and columns in the determinant of the integrand does not change the value of the determinant. Hence,

$$\overline{B_n}\begin{pmatrix} x_{11}, x_{12}, \ldots\ldots, x_{1p} \\ x_{21}, x_{22}, \ldots\ldots, x_{2p} \end{pmatrix}$$

$$= \int_a^b \ldots \int_a^b K\begin{pmatrix} x_{21}, x_{22}, \ldots\ldots, x_{2p}, t_1, t_2, \ldots\ldots, t_n \\ x_{11}, x_{12}, \ldots\ldots, x_{1p}, t_1, t_2, \ldots\ldots, t_n \end{pmatrix} dt_1 dt_2 \ldots dt_n$$

Then using the equation (3.6.1) defining B_n, we see that

$$\overline{B_n}\begin{pmatrix} x_{11}, x_{12}, \ldots\ldots, x_{1p} \\ x_{21}, x_{22}, \ldots\ldots, x_{2p} \end{pmatrix} = B_n\begin{pmatrix} x_{21}, x_{22}, \ldots\ldots, x_{2p} \\ x_{11}, x_{12}, \ldots\ldots, x_{1p} \end{pmatrix}$$

and therefore

https://doi.org/ 10.2478/9788366675575-003

$$\overline{D}\begin{pmatrix} x_{11}, x_{12}, \ldots\ldots, x_{1p} & \lambda \\ x_{21}, x_{22}, \ldots\ldots, x_{2p} & \end{pmatrix} = D\begin{pmatrix} x_{21}, x_{22}, \ldots\ldots, x_{2p} & \lambda \\ x_{11}, x_{12}, \ldots\ldots, x_{1p} & \end{pmatrix} \quad (3.8.4)$$

Since λ_0 is a characteristic constant of $K(x,t)$ of index q, we have

$$D\begin{pmatrix} x_{11}, x_{12}, \ldots\ldots, x_{1p} & \lambda_0 \\ x_{21}, x_{22}, \ldots\ldots, x_{2p} & \end{pmatrix} \equiv 0, \ (p = 1, 2, \ldots., q-1)$$

While

$$D\begin{pmatrix} x'_{11}, x'_{12}, \ldots\ldots, x'_{1p} & \lambda_0 \\ x'_{21}, x'_{22}, \ldots\ldots, x'_{2p} & \end{pmatrix} \neq 0$$

Hence by (3.8.4)

$$\overline{D}\begin{pmatrix} x_{11}, x_{12}, \ldots\ldots, x_{1p} & \lambda_0 \\ x_{21}, x_{22}, \ldots\ldots, x_{2p} & \end{pmatrix} = D\begin{pmatrix} x_{21}, x_{22}, \ldots\ldots, x_{2p} & \lambda_0 \\ x_{11}, x_{12}, \ldots\ldots, x_{1p} & \end{pmatrix} \equiv 0$$

In $x_{11}, x_{12}, \ldots\ldots, x_{1p}, x_{21}, x_{22}, \ldots\ldots, x_{2p}$ for $p = 1, 2, \ldots., q-1$.

Further, if we put

$$\overline{x'}_{1\alpha} = x'_{2\alpha}, \ \overline{x}'_{2\alpha} = x'_{1\alpha} \quad (3.8.5)$$

We have

$$\overline{D}\begin{pmatrix} \overline{x'}_{11}, \overline{x'_{12}}, \ldots\ldots, \overline{x'_{1q}} & \lambda_0 \\ \overline{x'}_{21}, \overline{x'_{22}}, \ldots\ldots, \overline{x'_{2q}} & \end{pmatrix} = D\begin{pmatrix} \overline{x'}_{21}, \overline{x'_{22}}, \ldots\ldots, \overline{x'_{2q}} & \lambda_0 \\ \overline{x'}_{11}, \overline{x'_{12}}, \ldots\ldots, \overline{x'_{1q}} & \end{pmatrix}$$

$$= D\begin{pmatrix} x'_{11}, x'_{12}, \ldots\ldots, x'_{1q} & \lambda_0 \\ x'_{21}, x'_{22}, \ldots\ldots, x'_{2q} & \end{pmatrix} \neq 0$$

https://doi.org/ 10.2478/9788366675575-003

Hence by definition, the index $\bar{q}$ of λ_0 as a characteristic constant of $K(t,x)$ is q. We state this result in the theorem.

Theorem-3.6

If λ_0 is a characteristic constant of $K(x,t)$ of index q, then λ_0 is a characteristic constant $K(t,x)$ of the same index:

$\bar{q} = q$. □

3.9. The Orthogonality Theorem:

In the previous two sections, we have considered Eigenvalues and Eigenfunctions of Fredholm equations of the second kind for any given kernel $K(x,t)$. In this section, we assume that the kernel $K(x,t)$ is symmetric, i.e., $K(x,t) = K(t,x)$.

Theorem-3.6

If $K(x,t)$ is real and symmetric, then the Eigenfunctions corresponding to distinct eigenvalues are orthogonal.

Proof:

Let λ_1 and λ_2 be two Eigenvalues and $y_1(x)$ and $y_2(x)$ be the corresponding Eigenfunctions of the equation

https://doi.org/ 10.2478/9788366675575-003

$$y(x)=\lambda\int_a^b K(x,t)y(t)dt \tag{3.9.1}$$

Then

$$y_1(x)=\lambda_1\int_a^b K(x,t)y_1(t)dt \tag{3.9.2}$$

$$y_2(x)=\lambda_2\int_a^b K(x,t)y_2(t)dt \tag{3.9.3}$$

Upon multiplying equation (3.9.2) by $y_2(x)$ and integrating the product from $x=a\ to\ b$. Thus, it obtains

$$\int_a^b y_1(x)y_2(x)dx=\lambda_1\int_a^b y_1(t)\left[\int_a^b K(x,t)y_2(x)dx\right]dt$$

$$=\lambda_1\int_a^b y_1(t)\left[\int_a^b K(t,x)y_2(x)dx\right]dt$$

$$=\frac{\lambda_1}{\lambda_2}\int_a^b y_1(t)y_2(t)dt$$

i.e. $(\lambda_1-\lambda_2)\int_a^b y_1(x)y_2(x)dx=0$

Since $\lambda_1\neq\lambda_2$, we find

$$\int_a^b y_1(x)y_2(x)dx=0$$

i.e., $y_1(x)$ and $y_2(x)$ are orthogonal in (a,b).

https://doi.org/ 10.2478/9788366675575-003

Definition:

Two continuous functions $y_1(x), y_2(x)$ for which

$$\int_a^b y_1(x)y_2(x)dx = 0$$

are said to be *orthogonal* to each other.

Hence the above result may be stated as follows: $y_1(x)$ and $y_2(x)$ are orthogonal to each other.

3.10. The Non-homogenous Integral equation when $D(\lambda) \equiv 0$:

With the aid of the results, we can discuss the solution of the non-homogenous integral equation completely

$$y(x) = f(x) + \lambda_0 \int_a^b K(x,t)y(t)dt \qquad (3.10.1)$$

When $D(\lambda_0) \equiv 0$ and λ_0 is of the index q.

The finite system of linear equations

$$y_i - \lambda h \sum_{j=1}^{n} K_{ij} y_j = f_i, \ (i = 1,2,....,n)$$

The equation (3.10.1) may be considered a limit, has in general for $\Delta = 0$, no finite solution.

https://doi.org/ 10.2478/9788366675575-003

If, however, certain conditions on f_i are satisfied, the system has an infinitude solution. In analogy, we will find that (3.10.1) has no solutions in general. If, however, $f(x)$ satisfies certain conditions, then (3.10.1) has an infinitude of solutions.

(a) Necessary Conditions:

To obtain these conditions $f(x)$,we assume that $y(x)$ is a continuous function of x satisfying (51).

Multiply both sides of (3.10.1) by $\overline{\phi_\alpha}(x)$,

Where $\overline{\phi_\alpha}(x)=\lambda_0\int_a^b K(t,x)\overline{\phi_\alpha}(t)dt$ and Integrate with respect to x from a to b , we obtain

$$\int_a^b f(x)\overline{\phi_\alpha}(x)dx$$
$$=\int_a^b y(x)\overline{\phi_\alpha}(x)dx-\lambda_0\int_a^b \overline{\phi_\alpha}(x)\left\{\int_a^b K(x,t)y(t)dt\right\}dx \qquad (3.10.2)$$

In the last integral on the right, $\overline{\phi_\alpha}(x)$ is constant concerning t and so can be placed under the second signs of integration.

We may change the order of integration and then take $y(t)$ from under the sign of integration concerning x .

Thus, the last term becomes

$$\int_a^b y(t)\left\{\lambda_0\int_a^b \overline{\phi_\alpha}(x)K(x,t)\,dx\right\}dt = \int_a^b y(t)\overline{\phi_\alpha}(t)\,dx$$

Thus, we see that the first and last terms on the right cancel and

$$\int_a^b f(x)\overline{\phi_\alpha}(x)\,dx = 0\ ,\ (\alpha = 1,2,\dots,q) \qquad (3.10.3)$$

Hence so that there may exist a continuous solution $y(x)$ of (3.10.1), $f(x)$ must satisfy the q conditions (3.10.3).

(b) Sufficient Proof:

Let us now show conversely that if $f(x)$ satisfies the q conditions (3.10.3) then (3.10.1) do have a solution.

By our hypothesis, the q equation (3.10.3) are satisfied. Then

$$\sum_{\alpha=1}^{q}\lambda_0 K(x,y'_\alpha)\int_a^b f(t)\overline{\phi_\alpha}(t)\,dt = 0$$

Now $\lambda_0 K(x,y'_\alpha)$, is independent of t and so may be placed under the sign of integration.

https://doi.org/ 10.2478/9788366675575-003

It is permissible to change the order of performing the summation and the integration.

We thus obtain $\int_a^b \left\{ \sum_{\alpha=1}^{q} \lambda_0 K(x, y'_\alpha) \overline{\phi_\alpha}(t) f(t) \right\} dt = 0$

Which on simplification, it becomes

$$0 = \int_a^b \lambda_0 K(x,t) f(t) dt - \int_a^b H(x,t) f(t) dt + \lambda_0 \int_a^b f(t) \left\{ \int_a^b H(s,t) K(x,s) ds \right\} dt \tag{3.10.4}$$

The last term may be written

$$\lambda_0 \int_a^b \int_a^b f(t) H(s,t) K(x,s) ds\, dt$$

Make now a change in notation. In place of t and s : write s and t, we obtain

$$\lambda_0 \int_a^b \int_a^b f(s) H(t,s) K(x,t) dt\, ds$$

In this definite double integral, it is permissible to change the order of integration.

We then obtain

$$\lambda_0 \int_a^b K(x,t) \left\{ \int_a^b H(t,s) f(s) ds \right\} dt$$

https://doi.org/ 10.2478/9788366675575-003

After making these reductions in (3.10.4), combine the first and last terms and obtain

$$0=\lambda_0\int_a^b K(x,t)\left\{f(t)+\int_a^b H(t,s)f(s)ds\right\}dt \\ -\int_a^b H(x,t)f(t)dt \tag{3.10.5}$$

Now put $y_0(t)=f(t)+\int_a^b H(t,s)f(s)ds$

Then $\int_a^b H(x,t)f(t)dt=y_0(x)-f(x)$

Making use of these last two equations (3.10.5) becomes

$$y_0(x)=f(x)+\lambda_0\int_a^b H(x,t)y_0(t)dt$$

Thus, we have proved that if (3.10.3) are satisfied, then (3.10.1) has at least one solution, $y_0(x)$ given by

$$y_0(x)=f(x)+\int_a^b H(x,t)f(t)dt \tag{3.10.6}$$

(c) Determination of all solutions:

Let us suppose that (3.10.1) has another continuous solution $y(x)$, then $y(x)-y_0(x)$ is a solution of the homogenous equation

https://doi.org/ 10.2478/9788366675575-003

$$z(x)=\lambda_0\int_a^b K(x,t)z(t)dt \tag{3.10.7}$$

For, if we subtract the members of (3.10.6) from the corresponding member of (3.10.1), we obtain

$$y(x)-y_0(x)=\lambda_0\int_a^b K(x,t)\left[y(t)-y_0(t)\right]dt \tag{3.10.8}$$

The most general solution of (3.10.7) is of the form

$C_1\phi_1(x)+C_2\phi_2(x)+.......+C_q\phi_q(x)$

Where $C_1, C_2, ..., C_q$ are arbitrary constants.

Hence the most general solution of (3.10.8) is

$y(x)-y_0(x)=C_1\phi_1(x)+C_2\phi_2(x)+.......+C_q\phi_q(x)$

Therefore

$$y(x)=f(x)+\int_a^b H(x,t)f(t)dt$$
$$+C_1\phi_1(x)+C_2\phi_2(x)+.......+C_q\phi_q(x)$$

is the complete solution of (3.10.1).

We have thus proved *Fredholm's third fundamental theorem*:

Theorem-3.7

If λ_0 is a characteristic constant of $K(x,t)$ of index q, then

$$y(x) = f(x) + \lambda_0 \int_a^b K(x,t) y(t) dt \tag{3.10.1}$$

has, in general, no continuous solution.

So that a continuous solution exists, it is necessary that

$$\int_a^b f(x) \overline{\phi_\alpha}(x) dx = 0 \text{ , } \alpha = 1, 2, \ldots\ldots, q$$

Where the $\overline{\phi_\alpha}(x)$ is a complete set of fundamental functions for the associated homogenous equation

$$z(x) = \lambda_0 \int_a^b K(x,t) z(t) dt \tag{3.10.2}$$

If these conditions are satisfied, then there are a $q-$ fold infinitude of solutions of (3.10.1) given by

$$y(x) = f(x) + \int_a^b H(x,t) f(t) dt$$
$$+ C_1\phi_1(x) + C_2\phi_2(x) + \ldots\ldots. + C_q\phi_q(x)$$

Where $C_1, C_2, \ldots\ldots., C_q$ are arbitrary constants and where the $\phi_\alpha(x)$ are a complete set of fundamental functions for

$$y(x) = \lambda_0 \int_a^b K(x,t) y(t) dt$$

and $H(x,t)$ is given by

https://doi.org/ 10.2478/9788366675575-003

$$H(x_1,x_2)=\frac{D\begin{pmatrix} x_1, x'_{11},\ldots\ldots, x'_{1q} & \lambda_0 \\ x_2, x'_{21},\ldots\ldots, x'_{2q} & \end{pmatrix}}{D\begin{pmatrix} x'_{11},\ldots\ldots, x'_{1q} & \lambda_0 \\ x'_{21},\ldots\ldots, x'_{2q} & \end{pmatrix}}.$$

The following table exhibits the results of the solution of Fredholm's equation together with the analogy between the finite system and the integral equation.

$$y(x)=f(x)+\lambda\int_a^b K(x,t)\,y(t)\,dt$$

Case-I: $D(\lambda)\neq 0$		**Case-II:** $D(\lambda)=0$ **,Index q**	
Non-Homogenous	Homogenous	Non-Homogenous	Homogenous
Unique Solution $y(x)=f(x)+\int_a^b \frac{D(x,t;\lambda)}{D(\lambda)} f(t)\,dt$	Unique solution $y\equiv 0$	In general, No Solution. Solution exists only if f satisfies $\int_a^b f(x)\overline{\phi_\alpha}(x)\,dx=0$ Then ∞^q solution.	∞^q Solution $\sum_{\alpha=1}^{q} c_\alpha \phi_\alpha(x)$

$$y_i - \lambda h \sum_{j=1}^{n} K_{ij} y_j = f_j, \ (i = 1, 2, ..., n)$$

Case-I: $\Delta \neq 0$		**Case-II:** $\Delta = 0$ **,Index** q	
Non-Homogenous	Homogenous	Non-Homogenous	Homogenous
Unique Solution $y_k = \frac{\sum_{i=1}^{n} f_i \Delta_{ik}}{\Delta}$	Unique Solution $y_k = 0$	In general, no solution ∞^q solutions if the f_i satisfy certain q relation of the form $c_{\alpha 1} f_1 + c_{\alpha 2} f_2 + ..$ $..... + c_{\alpha n} f_n = 0$ $, \alpha = 1, 2, ..., q$	∞^q solutions, if Δ is of rank r, where $q = n - r$.

3.11. Kernels of the form $\sum a_i(x) b_i(y)$:

We give a brief discussion of the integral equation with a kernel of the form

$$K(x, y) = a_1(x) b_1(y) + ... + a_n(x) b_n(y)$$

Fredholm's integral equation

$$y(x)=f(x)+\lambda\int_a^b K(x,t)\,y(t)\,dt$$

can now be written in the form

$$y(x)=f(x)+\lambda\left[a_1(x)\int_a^b b_1(t)\,y(t)\,dt+a_2(x)\int_a^b b_2(t)\,y(t)\,dt+........+a_n(x)\int_a^b b_n(t)\,y(t)\,dt\right] \quad (3.11.1)$$

If now we put

$$\int_a^b b_i(t)\,y(t)\,dt=K_i\,,\ (i=1,2,...,n) \quad (3.11.2)$$

We see that $y(x)$ is of the form

$$y(x)=f(x)+\lambda\left[a_1(x)K_1+.........+a_n(x)K_n\right] \quad (3.11.3)$$

To determine the constant K_i, let us substitute in (3.11.2), the value of y given by (3.11.3),

We obtain the n-equations

$$K_i-\lambda\left[\int_a^b a_1(t)b_i(t)K_1dt+\int_a^b a_2(t)b_i(t)K_2dt+.. ......+\int_a^b a_n(x)b_i(t)K_ndt\right] \quad (3.11.4)$$

$$=\int_a^b b_i(t)f(t)\,dt\ ,\ (i=1,2,...,n)$$

Introducing the notation

$$\int_a^b a_k(t)\,b_i(t)\,dt = c_{ki}$$

The system (3.11.4) can be written in the form

$$K_i - \lambda\left[c_{1i}K_1 + \dots\dots\dots + c_{ni}K_n\right] = \int_a^b b_i(t)\,f(t)\,dt \qquad (3.11.5)$$

Equation (3.11.5) is a linear algebraic system of n-non-homogenous equations in n-unknowns $K_1, K_2, \dots, K_n$, with the determinant

$$D(\lambda) = \begin{vmatrix} 1-\lambda c_{11} & -\lambda c_{12} & \cdots & -\lambda c_{1n} \\ -\lambda c_{21} & 1-\lambda c_{22} & \cdots & -\lambda c_{2n} \\ \cdots & \cdots & \vdots & \cdots \\ -\lambda c_{n1} & -\lambda c_{n2} & \cdots & 1-\lambda c_{nn} \end{vmatrix}$$

From the theory of linear system, we have at once the result:

(a) If $D(\lambda) \neq 0$, then the system (3.11.4) is satisfied by one set of values $K_1, K_2, \dots, K_n$ and these values are given by *Crammer's formula*. Therefore, Fredholm's equation (3.11.1) has one and only one solution, which is given by (3.11.3).

(b) If $D(\lambda) = 0$ for $\lambda = \lambda_0$ (and this happens for n-values of, real or complex), and one of the q th minors of $D(\lambda)$ is the

https://doi.org/ 10.2478/9788366675575-003

minor which does not vanish for $\lambda = \lambda_0$. (This q th minor is a determinant of order $n-q$), then the general solution of the homogenous system (3.11.4) ($f(x) \equiv 0$) will be of the form

$$K_i = \alpha_1 m_{1i} + \alpha_2 m_{2i} + ... + \alpha_q m_{qi}, \ (i = 1, 2, ..., n)$$

Where $\alpha_1, \alpha_2, ..., \alpha_q$ are arbitrary constants.

If we put the values of K_i so obtained in (3.11.3), we obtain

$$y(x) = \lambda_0 \left[\alpha_1 y_1(x) + \alpha_2 y_2(x) + ... + \alpha_q y_q(x) \right]$$

Where the functions

$$y_r(x) = m_{r1} a_1(x) + m_{r2} a_2(x) + ... + m_{rn} a_n(x), \quad (r = 1, 2, ..., q)$$

are linearly independent.

Thus, we see that, under the circumstances specified, the homogenous integral equation for $\lambda = \lambda_0$ has q linearly independent solutions.

The associated equation

https://doi.org/ 10.2478/9788366675575-003

$$\bar{y}(x) = f(x) + \lambda\left[b_1(x)\int_a^b a_1(t)\bar{y}(t)\,dt + b_2(x)\int_a^b a_2(t)\bar{y}(t)\,dt + \dots + b_n(x)\int_a^b a_n(t)\bar{y}(t)\,dt \right] \quad (3.11.6)$$

is obtained from (3.11.1) by interchanging the functions $a_i(x)$ and $b_i(x)$.

The general term of the characteristic determinant of (3.11.1) being

$$\int_a^b a_k(t)b_i(t)\,dt = c_{ki}$$

The general term for the associated equation will be

$$\int_a^b b_k(t)a_i(t)\,dt = c_{ik}\,.$$

The characteristic determinants of these two equations are identical since one can obtain one from the other by interchanging rows and columns.

Therefore, the equation (3.11.1) and the associated equation (3.11.6) have precisely the same characteristic numbers and with the same index.

https://doi.org/ 10.2478/9788366675575-003

From the general theory, we know that if λ_0 is a root of $D(\lambda)=0$ of index q, then, so that the non-homogenous equation (3.11.1) may have a solution, we must have

$$\int_a^b f(x)\overline{y_i}(x)dx=0,\ (i=1,2,....,q).\square$$

3.12. Eigen Values and Eigenfunctions

3.12.1Fredholm integral equation of the second kind

The homogenous Fredholm integral equation of the second kind

$$y(x)=\lambda\int_a^b K(x,t)y(t)dt \qquad (3.12.1)$$

always has the obvious solution $y(x)\equiv 0$, which is called the ***trivial solution***.

This equation's parameter values have nontrivial solutions called ***Eigenvalues*** or ***Characteristic numbers,*** and the ***nontrivial solutions*** are called ***Eigenfunctions***.

The number $\lambda=0$ is not an ***eigenvalue*** for if $\lambda=0$, then (3.12.1) yield the trivial solution, i.e., $y(x)=0$.

Find the eigenvalues and eigenfunctions for the following integral equations.

Example-1

$$y(x)=\lambda\int_0^{\frac{\pi}{4}}\sin^2 x\, y(t)\,dt$$

Solution:

Here $y(x)=\lambda c\sin^2 x$, where

$$c=\int_0^{\frac{\pi}{4}} y(t)\,dt$$

$$=\lambda c\int_0^{\frac{\pi}{4}}\sin^2 t dt$$

$$=\lambda c\frac{(\pi-2)}{8}$$

$\therefore \lambda=\dfrac{8}{\pi-2}$, is the eigenvalue;

Which implies $y(x)=\sin^2 x$ is the eigenfunction. □

Example-2

$$y(x)=\lambda\int_0^{2\pi}\sin x\cos t\, y(t)\,dt$$

https://doi.org/ 10.2478/9788366675575-003

Solution:

Here $y(x) = \lambda c \sin x$, where

$$c = \int_0^{2\pi} \cos t \, y(t)\, dt = \int_0^{2\pi} \lambda c \sin t \cos t \, dt = 0$$

Which implies that there is no Eigen value. □

Example-3

$$y(x) = \lambda \int_0^{2\pi} \sin x \sin t \, y(t)\, dt$$

Solution:

Here $y(x) = \lambda c \sin x$,

Where $c = \int_0^{2\pi} \lambda c \sin^2 t \, dt = \lambda c \pi$

It implies $\lambda = \frac{1}{\pi}$ be the Eigenvalue,

and the corresponding Eigenfunction be

$y(x) = \sin x$. □

Example-4

$$y(x) = \lambda \int_0^1 (3x-2)\, t \, y(t)\, dt$$

Solution:

Here $y(x)=\lambda(3x-2)c$,where

$$c=\int_0^1 \mathrm{t}\, y(t)\,dt=\lambda c=\int_0^1 \left(3t^2-2t\right)dt=0$$

Which implies that there is no Eigenvalue.

Example-5

$$y(x)=\lambda\int_0^1 \left(\sqrt{x}\,t-\sqrt{t}\,x\right)y(t)\,dt$$

Solution:

Here $y(x)=c_1\lambda\sqrt{x}-c_2\lambda x$, where

$$c_1=\int_0^1 \mathrm{t}\, y(t)\,dt=\int_0^1 t\left(c_1\lambda\sqrt{t}-c_2\lambda t\right)dt$$

$$c_2=\int_0^1 \sqrt{\mathrm{t}}\, y(t)\,dt=\int_0^1 \sqrt{t}\left(c_1\lambda\sqrt{t}-c_2\lambda t\right)dt$$

Evaluating the integrals, we get

$$\left(1-\frac{2\lambda}{5}\right)c_1+\frac{\lambda}{3}c_2=0$$

$$-\frac{\lambda}{2}c_1+\left(1+\frac{2\lambda}{5}\right)c_2=0$$

https://doi.org/ 10.2478/9788366675575-003

$$\begin{vmatrix} 1-\dfrac{2\lambda}{5} & \dfrac{\lambda}{3} \\ -\dfrac{\lambda}{2} & 1+\dfrac{2\lambda}{5} \end{vmatrix} = 1+\frac{\lambda^2}{150} \neq 0$$

$\therefore c_1 = c_1 = 0$, Which implies that there are no real eigenvalues. □

Example-6

$$y(x) = \lambda \int_{-1}^{1} (1+3xt)\, y(t)\, dt$$

Solution:

Here $y(x) = \lambda c_1 + 3\lambda c_2 x$

Where

$$c_1 = \int_{-1}^{1} y(t)\, dt$$

$$= \int_{-1}^{1} (\lambda c_1 + 3\lambda c_2 t)\, dt = 2\lambda c_1$$

$$c_2 = \int_{-1}^{1} t y(t)\, dt$$

$$= \int_{-1}^{1} \left(\lambda c_1 t + 3\lambda c_2 t^2\right) dt = 2\lambda c_2$$

https://doi.org/ 10.2478/9788366675575-003

Which gives $\lambda = \frac{1}{2}$, Eigen value.

Whereas $y(x) = 1, y(x) = x$ are the corresponding Eigenfunctions. □

Example-7

$$y(x) = \lambda \int_{-1}^{1} (1 + t + 3xt) y(t) dt$$

Solution:

Here $y(x) = \lambda c_1 + \lambda(1 + 3x) c_2$

Where,

$$c_1 = \int_{-1}^{1} y(t) dt$$

$$= \int_{-1}^{1} \left[\lambda c_1 + \lambda(1 + 3t) c_2\right] dt$$

Or $c_1 = 2\lambda(c_1 + c_2)$

$$c_2 = \int_{-1}^{1} t y(t) dt$$

$$= \int_{-1}^{1} \left[\lambda c_1 t + \lambda\left(t + 3t^{t}\right) c_2\right] dt = 2\lambda c_2$$

https://doi.org/ 10.2478/9788366675575-003

It implies $\lambda = \frac{1}{2}$ be the Eigenvalue, and the corresponding Eigenfunction is $y(x)=1$. □

Example-8

$$y(x)=\lambda\int_0^{2\pi}\sin(x+t)\,y(t)\,dt$$

Solution:

Here $y(x)=\lambda c_1\sin x+\lambda c_2\cos x$

Where

$$c_1=\int_0^{2\pi}\cos t\,y(t)\,dt$$

$$=\int_0^{2\pi}\left[\lambda c_1\,\text{sint}\cos t+\lambda c_2\cos^2 t\right]dt$$

$$\therefore c_1-\lambda\pi c_2=0$$

$$c_2=\int_0^{2\pi}\sin t\,y(t)\,dt$$

$$=\int_0^{2\pi}\left[\lambda c_1\sin^2\text{t}+\lambda c_2\,\text{sint}\cos t\right]dt$$

$$\therefore c_2-\lambda\pi c_1=0$$

Which implies

$$c_1 - \lambda\pi c_2 = 0$$
$$-\lambda\pi c_1 + c_2 = 0$$

$$\begin{vmatrix} 1 & -\lambda\pi \\ -\lambda\pi & 1 \end{vmatrix} = 1 - \lambda^2\pi^2 .$$

Thus, the eigenvalues are

$$\lambda_1 = \frac{1}{\pi} \text{ and } \lambda_2 = -\frac{1}{\pi} .$$

The corresponding eigenfunctions are

$y_1(x) = \sin x + \cos x$ and $y_2(x) = \sin x - \cos x$. □

Example-9

$$y(x) = \lambda \int_0^{2\pi} \cos(x+t) y(t) dt$$

Solution:

Here $y(x) = \lambda c_1 \cos x - \lambda c_2 \sin x$

Where

$$c_1 = \int_0^{2\pi} \cos \mathrm{t}\, y(t) dt$$
$$= \int_0^{2\pi} \left[\lambda c_1 \cos^2 t - \lambda c_2 \,\mathrm{sint}\,\mathrm{cost} \right] dt$$
$$= \pi\lambda c_1$$

$$\therefore c_1(1-\pi\lambda)=0$$

$$c_2=\int_0^{2\pi}\sin t\, y(t)\,dt$$

$$=\int_0^{2\pi}\left[\lambda c_1\cos t\sin t-\lambda c_2\sin^2 t\right]dt$$

$$=-\lambda\pi c_2$$

$$\therefore c_2(1+\lambda\pi)=0$$

The eigenvalues are

$$\lambda_1=\frac{1}{\pi} \text{ and } \lambda_2=-\frac{1}{\pi}$$

and the eigenfunctions are

$y_1(x)=\sin x$, $y_2(x)=\cos x$. □

Example-10

$$y(x)=\lambda\int_0^{\pi}\left(\cos^2 x\cos 2t+\cos 3x\cos^3 t\right)y(t)\,dt$$

Solution:

Here $y(x)=\lambda c_1\cos^2 x+\lambda c_2\cos 3x$

Where

https://doi.org/ 10.2478/9788366675575-003

$$c_1 = \int_0^{\pi} \cos 2t \, y(t)\, dt$$

$$= \int_0^{\pi} \cos 2t \left[\lambda c_1 \cos^2 t + \lambda c_2 \cos 3t \right] dt$$

$$= \frac{\lambda \pi}{4} c_1$$

$$\therefore \left(1 - \frac{\lambda \pi}{4}\right) c_1 = 0 \qquad (3.12.2)$$

$$c_2 = \int_0^{\pi} \cos^3 t \, y(t)\, dt$$

$$= \int_0^{\pi} \cos^3 t \left[\lambda c_1 \cos^2 t + \lambda c_2 \cos 3t \right] dt$$

$$= \frac{\lambda \pi}{8} c_2$$

$$\therefore \left(1 - \frac{\lambda \pi}{8}\right) c_2 = 0 \qquad (3.12.3)$$

From (3.12.2) and (3.12.3),

we find that the eigenvalues are

$\lambda_1 = \dfrac{4}{\pi}$ and $\lambda_2 = \dfrac{8}{\pi}$.

The corresponding eigenfunctions are

$y_1(x) = \cos^2 x$ and $y_2(x) = \cos 3x$. □

https://doi.org/ 10.2478/9788366675575-003

3.13. Eigenvalues for Non-separable kernels

In the previous section, we have considered finding eigenvalues and eigenfunctions for separable kernels. In this section, we deal with non-separable kernels. Here we reduce the Fredholm equation to a boundary value problem and find its eigenvalues and eigenfunctions. The following examples will illustrate the procedure.

Example-11

Find the Eigenvalues and Eigenfunction of

$$y(x)=\lambda\int_0^1 K(x,t)y(t)\,dt$$

Where $K(x,t)=\begin{cases} t(x-1), & t<x \\ x(t-1), & x<t \end{cases}$.

Solution:

Here $y(x)=\lambda\int_0^x t(x-1)y(t)\,dt+\lambda\int_x^1 x(t-1)y(t)\,dt$

Using Leibnitz rule for differentiating under integral sign, we find

$$\frac{dy}{dx}=\lambda\int_0^x ty(t)\,dt+\lambda\int_x^1 (t-1)y(t)\,dt$$

$$\frac{d^2y}{dx^2} = \lambda xy(x) - \lambda(x-1)y(x) = \lambda y(x)$$

$$\therefore \frac{d^2y}{dx^2} - \lambda y = 0,$$

which is the desired differential equation. The relevant boundary conditions are $y(0) = y(1) = 0$.

The boundary value problem

$$\frac{d^2y}{dx^2} - \lambda y = 0,$$

$$y(0) = y(1) = 0$$

has the Eigenvalues $\lambda_n = -n^2\pi^2$,

and the Eigenfunctions are

$y_n = \sin n\pi x, n = 1, 2, 3, \ldots$. □

Example-12

Find the Eigenvalues and Eigenfunctions of

$$y(x) = \lambda \int_0^{\pi} K(x,t)\, y(t)\, dt$$

Where $K(x,t) = \begin{cases} \cos t \sin x, & t \le x \\ \cos x \sin t, & x \le t \end{cases}$.

Solution:

https://doi.org/ 10.2478/9788366675575-003

Here $y(x)=\lambda\sin x\int_0^x \cos t\, y(t)\,dt+\lambda\cos x\int_x^\pi \sin t\, y(t)\,dt$

$$\frac{dy}{dx}=\lambda\cos x\int_0^x \cos t\, y(t)\,dt-\lambda\sin x\int_x^\pi \sin t\, y(t)\,dt$$

$$\frac{d^2y}{dx^2}=-\lambda\sin x\int_0^x \cos t\, y(t)\,dt-\lambda\cos x\int_x^\pi \sin t\, y(t)\,dt+\lambda y(x)$$

$$\therefore \frac{d^2y}{dx^2}+(1-\lambda)y=0$$

From the expression for $y(x)$ and $\frac{dy}{dx}$,

we find

$y(\pi)=0$ and $y'(0)=0$.

Now the boundary value problem becomes

$$\frac{d^2y}{dx^2}+(1-\lambda)y=0,$$

$$y(\pi)=0,\ y'(0)=0$$

We find the Eigenvalues as

$$\lambda_n=1-\left(n+\frac{1}{2}\right)^2$$

and the Eigenfunction as

https://doi.org/ 10.2478/9788366675575-003

$$y_n(x)=\cos\left(n+\frac{1}{2}\right)x \qquad \square$$

Example-13

Find the Eigenvalues and Eigenfunctions of

$$y(x)=\lambda\int_0^1 e^{-|x-t|}y(t)\,dt$$

Solution:

Here $y(x)=\lambda\int_0^x e^{t-x}\,y(t)\,dt+\lambda\int_x^1 e^{x-t}\,y(t)\,dt$

$$\frac{dy}{dx}=-\lambda\int_0^x e^{t-x}\,y(t)\,dt+\lambda\int_x^1 e^{x-t}\,y(t)\,dt$$

$$\frac{d^2y}{dx^2}=\lambda\int_0^x e^{t-x}\,y(t)\,dt+\lambda\int_x^1 e^{x-t}\,y(t)\,dt-2\lambda y(x)$$

These equations yield the boundary value problem

$$\frac{d^2y}{dx^2}+(2\lambda-1)y=0,$$
$$y'(0)-y(0)=0,$$
$$y'(1)+y(1)=0$$

This gives the Eigenvalues

$$\lambda_n=\frac{1+\mu^2{}_n}{2}$$

https://doi.org/ 10.2478/9788366675575-003

and Eigenfunctions

$y_n(x) = \sin \mu_n x + \mu_n \cos \mu_n x$

Where μ_n are the roots of the equation

$2\mu \cot \mu = \mu^2 - 1.$ □

3.14. Volterra Integral Equation

The equation

$$y(x) = f(x) + \lambda \int_a^x K(x,t)\, y(t)\, dt \qquad (3.14.1)$$

where $f(x)$ and $K(x,t)$ are known functions, $y(x)$ is the unknown function and λ is a parameter, is called ***Volterra integral equation of the second kind***.

If $f(x) = 0$, then equation (3.12.1) takes the form,

$$y(x) = \lambda \int_a^x K(x,t)\, y(t)\, dt \qquad (3.14.2)$$

is called the ***homogenous Volterra integral equation*** of the second kind.

The equation

$$\int_a^x K(x,t)\, y(t)\, dt = f(x) \qquad (3.14.3)$$

https://doi.org/ 10.2478/9788366675575-003

where $f(x)$ and $K(x,t)$ are known functions and $y(x)$ is an unknown function, is called ***Volterra integral equation of the first kind***.

The Volterra integral equation of the first kind can always be transformed into the *second kind's Volterra integral equation*.

Differentiating equation (3.12.3), we find

$$K(x,x)y(x)+\int_a^x \frac{\partial K(x,t)}{\partial x} y(t)dt = f'(x) \tag{3.14.4}$$

Hence we get the equation of the second kind

$$y(x) = -\int_a^x \frac{\frac{\partial K(x,t)}{\partial x}}{K(x,x)} y(t)dt + \frac{f'(x)}{K(x,x)} \tag{3.14.5}$$

provided $K(x,x) \neq 0$.

If $K(x,x)=0$, then we differentiate equation (3.14.4) once again to get the equation of the second kind. □

Example-14

Reduce the following equation of the first kind into an equation of the second kind

$$\int_0^x e^{x-t} y(t)\,dt = \sin x$$

Solution:

Differentiating this equation, we get

$$y(x) + \int_0^x e^{x-t} y(t)\,dt = \cos x$$

□

Example-15

Reduce the equation into an equation of the second kind.

$$\int_0^x \cos(x-t)\,y(t)\,dt = x$$

Solution:

Differentiating, we get

$$y(x) - \int_0^x \sin(x-t)\,y(t)\,dt = 1$$

□

$$\Rightarrow y(x) = 1 + \int_0^x \sin(x-t)\,y(t)\,dt$$

Example-16

Reduce the equation into an equation of the second kind.

$$\int_0^x \sin(x-t)\,y(t)\,dt = e^{\frac{x^2}{2}} - 1$$

https://doi.org/ 10.2478/9788366675575-003

Solution:

Differentiating, we get

$$\int_0^x \cos(x-t)\,y(t)\,dt = xe^{\frac{x^2}{2}}$$

Differentiating again, we have

$$y(x) = \left(1+x^2\right)e^{\frac{x^2}{2}} + \int_0^x \sin(x-t)\,y(t)\,dt$$

Unlike the Fredholm integral equation of the second kind, the homogenous Volterra integral equation of the second kind does not possess eigenvalues. □

Theorem-3.8

For any λ, the equation

$$y(x) = \lambda\int_0^x K(x,t)\,y(t)\,dt \quad , \qquad (7.1.6)$$

has only the trivial solution.

Proof:

Let $|y(x)| < A$ and $K(x,t) \le M$

Then equation (7.1.6) implies

$$|y(x)| \le |\lambda| AMx$$

Substituting this estimate on the right side of (7.1.6),

https://doi.org/ 10.2478/9788366675575-003

we get

$$|y(x)| \le |\lambda|^2 M^2 \frac{Ax^2}{2!}$$

Continuing this process, we get

$$|y(x)| \le |\lambda|^n M^n \frac{Ax^n}{n!}$$

As this last expression tends to 0 for any λ ,

So, we find $y(x)=0$. □

3.15. Solution by Resolvent Kernel

Suppose we have a Volterra integral equation of the second kind,

$$y(x)=f(x)+\lambda\int_0^x K(x,t)y(t)dt \qquad (3.15.1)$$

We seek the solution of the equation of (3.15.1) in the form of an infinite series in λ.

$$y(x)=y_0(x)+\lambda y_1(x)+\lambda^2 y_2(x)+.....+\lambda^n y_n(x)+... \quad (3.15.2)$$

Substituting equation (3.15.2) in equation (3.15.1) and comparing the like powers of λ, we find

https://doi.org/ 10.2478/9788366675575-003

$$y_0(x) = f(x)$$

$$y_1(x) = \int_0^x K(x,t)\, y_0(t)\, dt = \int_0^x K(x,t)\, f(t)\, dt$$

$$y_2(x) = \int_0^x K(x,t)\, y_1(t)\, dt = \int_0^x K(x,t) \int_0^t K(\mathrm{t},t_1) f(t_1)\, dt$$

$$\vdots$$

This relation yields a method of successive differentiation of the functions $y_n(x)$.

By changing the order of integration, we find

$$y_2(x) = \int_0^x K_2(x,\mathrm{s})\, f(s)\, dt$$

Where $K_2(x,\mathrm{s}) = \int_0^x K(x,t) K(\mathrm{t},\mathrm{s})\, dt$

Or $y_2(x) = \int_0^x K_2(x,\mathrm{t})\, f(t)\, dt$.

Where, $K_2(x,\mathrm{t}) = \int_t^x K(x,\mathrm{s}) K(\mathrm{s},\mathrm{t})\, dt$.

In general,

$$y_n(x) = \int_0^x K_n(x,\mathrm{t})\, f(t)\, dt \tag{3.15.3}$$

https://doi.org/ 10.2478/9788366675575-003

Where $K_1(x,\mathrm{t}), K_2(x,\mathrm{t}),........, K_n(x,\mathrm{t})$ are called the iterated kernels are given by

$$K_1(x,\mathrm{t}) = K(x,t)$$

$$\therefore K_n(x,\mathrm{t}) = \int_t^x K_{n-1}(x,\mathrm{s})K(\mathrm{s},\mathrm{t})\,dt$$

Using equation (7.2.3) in (7.2.2), we find the solution

$$y(x) = f(x) + \sum_{n=1}^{\infty} \lambda^n \int_0^x K_n(x,t) f(t)\,dt$$

If we define the resolvent kernel as

$$R(x,t;\lambda) = \sum_{n=1}^{\infty} \lambda^n K_{n+1}(x,t)$$
$$= K(x,t) + \lambda K_2(x,t) + \lambda^2 K_3(x,t) + ...$$

Then the required solution is given by

$$y(x) = f(x) + \lambda \int_0^x R(x,t;\lambda) f(t)\,dt$$

Thus, the solution is obtained by first finding the iterated kernels $K_2(x,t), K_3(x,t),...$ and then constructing the resolvent kernel $R(x,t;\lambda)$. □

https://doi.org/ 10.2478/9788366675575-003

Example-4

Solve $y(x)=f(x)+\lambda\int_0^x y(t)\,dt$

Solution:

Here

$$K(x,t)=1$$

$$K_2(x,t)=\int_t^x 1\times 1\,ds=x-t$$

$$K_3(x,t)=\int_t^x (x-s)\,ds=\frac{(x-t)^2}{2!}$$

$$\vdots$$

$$K_n(x,t)=\frac{(x-t)^n}{\text{n!}}$$

Therefore, the resolvent kernel is given by

$$R(x,t;\lambda)=K(x,t)+\lambda K_2(x,t)+\lambda^2 K_3(x,t)+\ldots..$$

$$=1+\lambda(x-t)+\lambda^2\frac{(x-t)^2}{2!}+\ldots\ldots..$$

$$=e^{\lambda(x-t)}$$

Hence the solution is

$$y(x)=f(x)+\lambda\int_0^x e^{\lambda(x-t)}f(t)\,dt. \qquad \square$$

https://doi.org/ 10.2478/9788366675575-003

Example-5

Solve $y(x)=f(x)+\int_0^x (x-t)\,y(t)\,dt$

Solution:

Here

$K(x,t)=x-t$

$K_2(x,t)=\int_t^x (x-s)\times(s-t)\,ds=\frac{(x-t)^3}{3!}$

$K_3(x,t)=\int_t^x \frac{(x-s)^3}{3!}(s-t)\,ds=\frac{(x-t)^5}{5!}$

⋮

Since $\lambda=1$,

$$\begin{aligned}R(x,t;\lambda)&=K(x,t)+\lambda K_2(x,t)+\lambda^2K_3(x,t)+.....\\&=(x-t)+\frac{(x-t)^3}{3!}+........\\&=\sinh(x-t)\end{aligned}$$

Hence the solution is

$y(x)=f(x)+\lambda\int_0^x \sinh(x-t)\,f(t)\,dt$ □

https://doi.org/ 10.2478/9788366675575-003

Example-6

Solve $y(x)=e^{x^2}+\int_0^x e^{x^2-t^2}y(t)\,dt$

Solution:

Here $\lambda=1$,

$$K(x,t)=e^{x^2-t^2}$$

$$K_2(x,t)=\int_t^x e^{x^2-s^2}\times e^{s^2-t^2}\,ds=(x-t)e^{x^2-t^2}$$

$$K_3(x,t)=\int_t^x (x-s)e^{x^2-s^2}\times e^{s^2-t^2}\,ds=\frac{(x-t)^2}{2!}e^{x^2-t^2}$$

$$\vdots$$

$$K_n(x,t)=\frac{(x-t)^{n-1}}{(n-1)!}e^{x^2-t^2}$$

$$\therefore R(x,t;\lambda)=K(x,t)+\lambda K_2(x,t)+\lambda^2K_3(x,t)+.....$$

$$=e^{x^2-t^2}\left[1+(x-t)+\frac{(x-t)^2}{2!}+........\right]$$

$$=e^{x^2-t^2}e^{(x-t)}$$

Therefore, the solution is

$$y(x)=e^{x^2}+\int_0^x e^{x^2-t^2}e^{(x-t)}e^{t^2}\,dt$$

$$=e^{x^2}+e^{x+x^2}\left(1-e^{-x}\right)$$

$$=e^{x+x^2}$$ □

https://doi.org/ 10.2478/9788366675575-003

Example-7

Solve $y(x)=e^{x}+\int_{0}^{x}e^{x-t}y(t)\,dt$

Solution:

Here $\lambda=1$,

$$K(x,t)=e^{x-t}$$

$$K_2(x,t)=\int_{t}^{x}e^{x-s}\times e^{s-t}\,ds=(x-t)e^{x-t}$$

$$K_3(x,t)=\int_{t}^{x}(x-s)e^{x-s}\times e^{s-t}\,ds=\frac{(x-t)^2}{2!}e^{x-t}$$

$\vdots$

$$K_n(x,t)=\frac{(x-t)^{n-1}}{(n-1)!}e^{x-t}$$

$$\therefore R(x,t;\lambda)=K(x,t)+\lambda K_2(x,t)+\lambda^2K_3(x,t)+.....$$

$$=e^{(x-t)}\left[1+(x-t)+\frac{(x-t)^2}{2!}+........\right]$$

$$=e^{2(x-t)}$$

Thus, the solution is

$$y(x)=e^{x}+\int_{0}^{x}e^{2(x-t)}e^{t}\,dt$$

$$=e^{x}+e^{2x}\left(1-e^{-x}\right)$$

$$=e^{2x}$$

□

https://doi.org/ 10.2478/9788366675575-003

Example-8

Solve $y(x)=\sin x+2\int_0^x e^{x-t}y(t)\,dt$

Solution:

Here $R(x,t;\lambda)=e^{(x-t)}e^{2(x-t)}=e^{3(x-t)}$

$\therefore$ The solution is

$$y(x)=\sin x+2\int_0^x e^{3(x-t)}\sin t\,dt$$

$$=\frac{2}{5}\sin x-\frac{1}{5}\cos x+\frac{1}{5}e^{3x}$$

□

Example-9

Solve $y(x)=x.3^x-\int_0^x 3^{x-t}y(t)\,dt$

Solution:

Here $\lambda=-1$,

$$K(x,t)=3^{x-t}$$

$$K_2(x,t)=\int_t^x 3^{x-s}\times 3^{s-t}\,ds=(x-t)3^{x-t}$$

$$K_3(x,t)=\frac{(x-t)^2}{2!}3^{x-t}$$

etc.

$$\therefore R(x,t;\lambda) = K(x,t) + \lambda K_2(x,t) + \lambda^2 K_3(x,t) + ...$$

$$= 3^{(x-t)}\left[1-(x-t)-\frac{(x-t)^2}{2!}+........\right]$$

$$= 3^{(x-t)}e^{-(x-t)}$$

Thus, the solution is given by

$$y(x) = x.3^x - \int_0^x 3^{x-t}e^{-(x-t)}t.3^t\,dt$$

$$= x.3^x - 3^x e^{-x}\int_0^x t.\mathrm{e}^t\,dt$$

$$= 3^x\left(1-e^{-x}\right)$$

□

Example-10

Solve $y(x) = \mathrm{e}^x \sin x + \int_0^x \frac{2+\cos x}{2+\cos t}y(t)\,dt$

Solution:

Here $\lambda = 1$,

$$K(x,t) = \frac{2+\cos x}{2+\cos t}$$

$$K_2(x,t) = \int_t^x \frac{2+\cos x}{2+\cos s}\times\frac{2+\cos s}{2+\cos t}\,ds = (x-t)\left[\frac{2+\cos x}{2+\cos t}\right]$$

$$K_3(x,t) = \frac{(x-t)^2}{2!}\left[\frac{2+\cos x}{2+\cos t}\right]$$

https://doi.org/ 10.2478/9788366675575-003

$$\therefore R(x,t;\lambda) = K(x,t) + \lambda K_2(x,t) + \lambda^2 K_3(x,t) + ...$$

$$= \frac{2+\cos x}{2+\cos t}\left[1-(x-t)-\frac{(x-t)^2}{2!}+........\right]$$

$$= \frac{2+\cos x}{2+\cos t}e^{x-t}$$

Therefore, the solution is

$$y(x) = \mathrm{e}^x \sin x + \int_0^x \frac{2+\cos x}{2+\cos t} e^{x-t} e^t \sin t dt$$

$$= \mathrm{e}^x \sin x + (2+\cos x)e^x \int_0^x \frac{\sin t}{2+\cos t} dt$$ □

$$= \mathrm{e}^x \sin x + (2+\cos x)e^x \ln\left(\frac{3}{2+\mathrm{cosx}}\right)$$

Example-11

Solve $y(x) = 1-2x-\int_0^x e^{x^2-t^2} y(t) dt$

Solution:

Here $\lambda = -1$,

$K(x,t) = e^{x^2-t^2}$

$\therefore R(x,t;\lambda) = e^{x^2-t^2}.e^{-(x-t)}$

Thus, the solution is given by

https://doi.org/ 10.2478/9788366675575-003

$$y(x)=1-2x-\int_0^x e^{x^2-t^2}e^{t-x}(1-2t)\,dt$$

$$=1-2x-e^{x^2-x}\int_0^x e^{t-t^2}(1-2t)\,dt$$

$$=1-2x-e^{x^2-x}\left(e^{x-x^2}-1\right)$$

$$=e^{x^2-x}-2x$$

□

Example-12

Solve $y(x)=e^{x^2+2x}+2\int_0^x e^{x^2-t^2}y(t)\,dt$

Solution:

Here $\lambda=2$, $K(x,t)=e^{x^2-t^2}$

$R(x,t;\lambda)=e^{x^2-t^2}.e^{2(x-t)}$

Thus, the solution is given by

$$y(x)=e^{x^2+2x}+2\int_0^x e^{x^2-t^2}.e^{2(x-t)}e^{t^2+2t}\,dt$$

$$=e^{x^2+2x}+2xe^{x^2+2x}$$

$$=e^{x^2+2x}(1+2x)$$

□

Example-13

Solve $y(x)=1+x^2+2\int_0^x \frac{1+x^2}{1+t^2}.y(t)\,dt$

https://doi.org/ 10.2478/9788366675575-003

Solution:

Here $\lambda = 1$,

$$K(x,t) = \frac{1+x^2}{1+t^2}$$

$$K_2(x,t) = \int_t^x \frac{1+x^2}{1+s^2} \times \frac{1+s^2}{1+t^2} ds = (x-t)\left[\frac{1+x^2}{1+t^2}\right]$$

Similarly,

$$K_3(x,t) = \frac{(x-t)^2}{2!}\frac{1+x^2}{1+t^2}$$

$$R(x,t;\lambda) = K(x,t) + \lambda K_2(x,t) + \lambda^2 K_3(x,t) + ...$$

$$= \frac{1+x^2}{1+t^2}\left[1+(x-t)+\frac{(x-t)^2}{2!}+........\right]$$

$$= \frac{1+x^2}{1+t^2}e^{x-t}$$

Thus, the solution is given by

$$y(x) = 1+x^2+\int_0^x \frac{1+x^2}{1+t^2}.e^{x-t}\left(1+t^2\right)dt$$

$$= 1+x^2+\left(1+x^2\right)e^x\left(1-e^{-x}\right)$$ □

$$= e^x\left(1+x^2\right)$$

3.16. Method of Successive approximation

Consider the Volterra integral equation of the second kind

https://doi.org/ 10.2478/9788366675575-003

$$y(x) = f(x) + \lambda\int_0^x K(x,t)\,y(t)\,dt \tag{3.16.1}$$

We assume that the functions f and K are constants.

Let us take some initial function $y_0(x)$. Put it into the right side of equation (3.16.1) in the place of $y(x)$, we get

$$y_1(x) = f(x) + \lambda\int_0^x K(x,t)\,y_0(t)\,dt$$

$$y_2(x) = f(x) + \lambda\int_0^x K(x,t)\,y_1(t)\,dt$$

Proceeding like this, we get a sequence of functions

$y_0(x), y_1(x), y_2(x), \ldots, y_n(x), \ldots$

Where

$$y_n(x) = f(x) + \lambda\int_0^x K(x,t)\,y_{n-1}(t)\,dt$$

The sequence $\{y_n(x)\}$ converges to the desired solution of equation (3.16.1) as $n \to \infty$.

Generally, the function $f(x)$ is chosen as $y_0(x)$. But in some cases, the choice for $y_0(x)$ other than $f(x)$ may accelerate the convergence. □

https://doi.org/ 10.2478/9788366675575-003

Example-14

Solve $y(x)=1+\int_0^x y(t)\,dt$

Solution:

Taking $y_0(x)=1$, we get the succession,

$$y_1(x)=1+\int_0^x 1.dt=1+x$$

$$y_2(x)=1+\int_0^x (1+t).dt=1+x+\frac{x^2}{2!}$$

$$y_3(x)=1+\int_0^x \left(1+t+\frac{t^2}{2!}\right).dt=1+x+\frac{x^2}{2!}+\frac{x^3}{3!}$$

Likewise, we get

$$y_n(x)=1+x+\frac{x^2}{2!}+\frac{x^3}{3!}+......+\frac{x^n}{n!}$$

and $y(x)=\lim_{n\to\infty} y_n(x)=e^x$ .which is the desired solution. □

Example-15

Solve $y(x)=e^x+\int_0^x e^{x-t}y(t)\,dt$

Solution:

Taking $y_0(x)=e^x$,we get

https://doi.org/ 10.2478/9788366675575-003

$$y_1(x) = e^x + \int_0^x e^{x-t}.e^t dt$$

$$= e^x(1+x)$$

$$y_2(x) = e^x + \int_0^x e^{x-t}.e^t(1+t).dt$$

$$= e^x\left(1+x+\frac{x^2}{2!}\right)$$

Proceeding further, we get

$$y_n(x) = e^x\left(1+x+\frac{x^2}{2!}+\frac{x^3}{3!}+......+\frac{x^n}{n!}\right)$$

$$\therefore y(x) = \lim_{n\to\infty} y_n(x) = e^x.e^x = e^{2x}$$

Which is the desired solution. □

Example-16

Solve $y(x) = x.2^x - \int_0^x 2^{x-t} y(t)dt$

Solution:

Taking $y_0(x) = x.2^x$, we get

https://doi.org/ 10.2478/9788366675575-003

$$y_1(x) = x.2^x - \int_0^x 2^{x-t}.\text{t}.2^t\,dt$$

$$= 2^x\left(x - \frac{x^2}{2!}\right)$$

$$y_2(x) = 2^x\left(x - \frac{x^2}{2!} + \frac{x^3}{3!}\right)$$

$$\vdots$$

$$y_n(x) = 2^x\left(x - \frac{x^2}{2!} + \frac{x^3}{3!} - \ldots\ldots + (-1)^{n-1}\frac{x^n}{\text{n!}}\right)$$

$$\therefore y(x) = \lim_{n\to\infty} y_n(x) = 2^x\left(1 - e^{-x}\right)$$

Which is the required solution. □

Example-17

Solve: $y(x) = e^{x^2} + \int_0^x e^{x^2-t^2} y(t)\,dt$

Solution:

Taking $y_0(x) = e^{x^2}$, we get

https://doi.org/ 10.2478/9788366675575-003

$$y_1(x)=e^{x^2}+\int_0^x e^{x^2-t^2}.e^{t^2}\,dt=e^{x^2}(1+x)$$

$$y_2(x)=e^{x^2}\left(1+x+\frac{x^2}{2!}\right)$$

$$\vdots$$

$$y_n(x)=e^{x^2}\left(1+x+\frac{x^2}{2!}+\frac{x^3}{3!}+......+\frac{x^n}{n!}\right)$$

Taking the Limit as $n\to\infty$, we get the desired solution $y(x)=\lim_{n\to\infty}y_n(x)=e^{x^2+x}$. □

Example-18

Solve $y(x)=x-\int_0^x (x-t)y(t)\,dt$

Solution:

Taking $y_0(x)=x$, we get

$$y_1(x)=x-\int_0^x (xt-t^2).dt$$

$$=x-\left(\frac{x^3}{2}-\frac{x^3}{3}\right)$$

$$=x-\frac{x^3}{3!}$$

https://doi.org/ 10.2478/9788366675575-003

$$y_2(x) = x - \frac{x^3}{3!} + \frac{x^5}{5!}$$

$$\vdots$$

$$y_n(x) = x - \frac{x^3}{3!} + \frac{x^5}{5!} - ......+(-1)^n \frac{x^{2n+1}}{(2n+1)!}$$

Taking the Limit as $n \to \infty$ we find the solution $y(x) = \lim_{n \to \infty} y_n(x) = \sin x$. □

https://doi.org/ 10.2478/9788366675575-003

Exercises

1. Solve the linear Fredholm integral equation,

$$y(x) = x + \int_0^1 (1 + x + t)\, y(t)\, dt$$

2. Solve the linear Fredholm integral equation,

$$y(x) = x + \int_0^1 (x - t)\, y(t)\, dt$$

3. Solve the linear Fredholm integral equation,

$$y(x) = x + \int_0^1 (x - t)^2\, y(t)\, dt$$

4. Solve the linear Volterra Integral Equation,

$$y(x) = x + \int_0^x (t - x)\, y(t)\, dt \quad ,0 < x < 1$$

with initial condition $y(0) = 0$.

5. Solve the linear Volterra-Integral equation

$$y(x) = x + \int_0^x t y(t)\, dt$$

with initial condition $y(0) = 0$.

6. Solve the linear Volterra-Integral Equation

https://doi.org/ 10.2478/9788366675575-003

$$y(x)=1+2\sin x-\int_0^x y(t)\,dt$$

with initial condition $y(0)=1$

7. Solve the first kind Volterra Integral Equation

$$x^2+\frac{1}{6}x^3=\int_0^x (2+x-t)\,y(t)\,dt$$

8. Solve the first kind Volterra Integral Equation

$$xe^x=\int_0^x e^{x-t}y(t)\,dt$$

$$e^x=\int_0^x e^{-t}y(t)\,dt$$

subject to the initial condition $y(0)=1$

9. Solve the first kind Volterra Integral Equation

$$2e^x-x-2=\int_0^x (1+x-t)\,y(t)\,dt$$

subject to the initial condition $y(0)=1$

10. Solve the first kind Volterra Integral Equation

$$4\sin x-3\cos x+3=\int_0^x (4+3x-3t)\,y(t)\,dt$$

subject to the initial condition $y(0)=1$

https://doi.org/ 10.2478/9788366675575-003

11. Solve the first kind Volterra Integral Equation

$$xe^{-x} = \int_0^x e^{t-x} y(t)\,dt$$

$$x = \int_0^x e^t y(t)\,dt$$

subject to the initial condition $y(0) = 1$.

12. Solve the first kind Volterra Integral Equation

$$5x^2 + x^3 = \int_0^x (5 + 3x - 3t)\,y(t)\,dt$$

subject to the initial condition $y(0) = 0$

13. Consider first-order Volterra Integral Equation

$$\frac{e^{2x}}{2} - \frac{1}{2} = \int_0^x y^2(t)\,dt$$

subject to the initial condition

$y(0) = \pm 1$

Find the eigen values and eigen functions of the following integral equations.

14. $y(x) = \lambda \int_0^1 \left(45x^2 \ln t - 9t^2 \ln x\right) y(t)\,dt$

Ans. There are no real eigen values.

https://doi.org/ 10.2478/9788366675575-003

15. $y(x)=\lambda\int_{0}^{1}\left(2xt-4x^{2}\right)y(t)\,dt$

Ans. $\lambda=-3, y(x)=x-2x^{2}$

16. $y(x)=\lambda\int_{-1}^{1}\left(5xt^{3}+4x^{2}t\right)y(t)\,dt$

Ans. $\lambda=\frac{1}{2}, y(x)=\frac{5x}{2}+\frac{10}{3}x^{2}$

17. $y(x)=\lambda\int_{-1}^{1}\left(5xt^{3}+4x^{2}t+3xt\right)y(t)\,dt$

Ans. $\lambda=\frac{1}{4}, y(x)=\frac{3x}{2}+x^{2}$

18. $y(x)=\lambda\int_{-1}^{1}\left(x\cosh t-t\sinh x\right)y(t)\,dt$

Ans. $\lambda=-\frac{e}{2}, y(x)=\sinh x$

19. $y(x)=\lambda\int_{-1}^{1}\left(x\cosh t-t^{2}\sinh x\right)y(t)\,dt$

Ans. No eigen values.

Find the eigen values and eigen functions of the following integral equations

$y(x)=\lambda\int_{a}^{b}K(x,t)y(t)\,dt$,

https://doi.org/ 10.2478/9788366675575-003

Where

20. $$K(x,t) = (x-2)(t+1), t \le x \le 1$$
$$= (t-2)(x+1), 0 \le x \le t$$

Ans. $$\lambda_n = \frac{-\mu_n^2}{3}, y_n(x) = \sin \mu_n x + \mu_n \cos \mu_n x,$$

Where μ_n *are the roots of* $2\mu \cot \mu = \mu^2 - 1.$

21. $$K(x,t) = \sin t \cos x, t \le x \le \frac{\pi}{2}$$
$$= \text{sinx cost}, 0 \le x \le t$$

Ans. $\lambda_n = 4n^2 - 1, y_n(x) = \sin 2nx$

22. $$K(x,t) = \sin t \cos x, t \le x \le \pi$$
$$= \text{sinx cost}, 0 \le x \le t$$

Ans. $\lambda_n = \left(n + \frac{1}{2}\right)^2 - 1, y_n(x) = \sin\left(n + \frac{1}{2}\right)x$

23. $$K(x,t) = \sin t \sin(x-1), t \le x \le \pi$$
$$= \text{sinx} \sin(t-1), 0 \le x \le t$$

Ans. $$\lambda_n = \frac{1-\mu_n^2}{\sin 1}, y_n(x) = \sin \mu_n(\pi + x),$$

Where μ_n *are the roots of* $\tan 2\pi\mu + \mu \tan 1 = 0.$

https://doi.org/ 10.2478/9788366675575-003

24.

$$K(x,t)=\sin\left(t+\frac{\pi}{4}\right)\sin\left(x-\frac{\pi}{4}\right), t\le x\le \pi$$

$$=\sin\left(x+\frac{\pi}{4}\right)\sin\left(t-\frac{\pi}{4}\right), 0\le x\le t$$

Ans. $\lambda_n = 1-\mu_n^2, y_n(x)=\sin\mu_n x+\mu_n\cos\mu_n x,$
Where μ_n *are the roots of* $2\mu\cot\mu=\mu^2-1$

Solve the following Volterra integral equations.

25. $y(x)=e^{2x}+\int_0^x e^{t-x}y(t)\,dt$

Ans. $y(x)=\frac{1}{2}\left(3e^{2x}-1\right)$

26. $y(x)=x-\int_0^x (x-t)\,y(t)\,dt$

Ans. $y(x)=\sin x$

27. $y(x)=1+x+\int_0^x e^{-2(x-t)}y(t)\,dt$

Ans. $y(x)=2x+1$

28. $y(x)=\sin x+\int_0^x (x-t)\,y(t)\,dt$

Ans. $y(x)=\frac{1}{2}(\sin x+\sinh x)$

https://doi.org/ 10.2478/9788366675575-003

29. $y(x)=\cos x+\int_0^x y(t)\,dt$

Ans. $y(x)=\frac{1}{2}\left(e^x+\cos x+\sin x\right)$

30. $y(x)=\frac{1}{1+x^2}+\int_0^x \sin(x-t)\,y(t)\,dt$

Ans. $y(x)=\frac{1}{1+x^2}+x\tan^{-1}x-\frac{1}{2}\ln\left(1+x^2\right)$

Solve the following integral equations by the method of successive approximations.

31. $y(x)=1-\int_0^x (x-t)\,y(t)\,dt$

Ans. $y(x)=\cos x$

32. $y(x)=1+\int_0^x (x-t)\,y(t)\,dt$

Ans. $y(x)=\cosh x$

33. $y(x)=x+1-\int_0^x y(t)\,dt$

Take

$(a)\; y_0(x)=1,$

$(b)\; y_0(x)=x+1$

https://doi.org/ 10.2478/9788366675575-003

Ans. $y(x)=1$

34. $y(x)=x+1-\int_0^x y(t)\,dt$

Take

$(a)\; y_0(x)=1,$

(b) $y_0(x)=x,$

(c) $y_0(x)=x+\dfrac{x^2}{2}$

Ans. $y(x)=x$

35. $y(x)=1+x+\int_0^x (x-t)\,y(t)\,dt$

Take $y_0(x)=1$

Ans. $y(x)=e^x$

36. $y(x)=2+2x-\int_0^x y(t)\,dt$

Take

$(a)\; y_0(x)=1,$

(b) $y_0(x)=2$

Ans. $y(x)=2$

https://doi.org/ 10.2478/9788366675575-004

Chapter-4

Applications of Fredholm Theory

4.1. Free vibration of an elastic string:

4.1.1. The differential equation of the problem:

Let us consider an elastic string stretched between the two fixed points A and B. We pull it out of this position of equilibrium ***AQB*** into some other plane initial position, as ***ACB*** and then release it. The string will describe transverse vibrations.

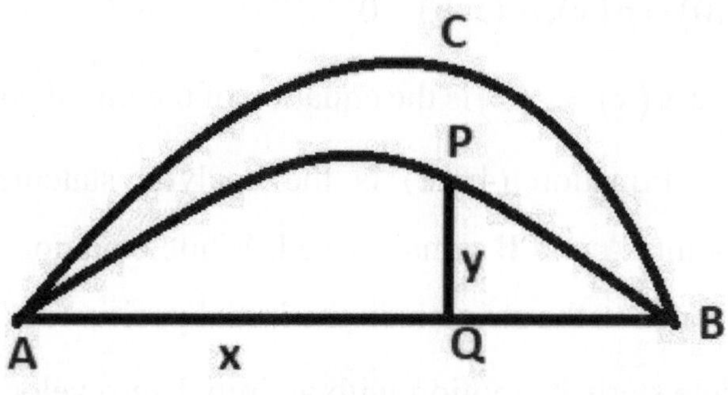

Fig.-4

Suppose that at the time, t the string occupies the position ***APB***. Let $x = AQ, y = QP$ be the abscissa and ordinate of any one of its points P. Then u is a function of x and t.

https://doi.org/ 10.2478/9788366675575-004

We suppose the cross-section of the string to be constant and infinitesimal compared with the length. The string is of homogenous density. The effect of gravity is to be neglected. Further, we take for simplicity. It is then proved in the theory of elasticity that the partial differential equation gives the motion of the string

$$\frac{\partial^2 u}{\partial t^2} = c^2 \frac{\partial^2 u}{\partial x^2} \quad (c = \text{constant}) \tag{4.1.1}$$

With initial conditions

$$u(0,t) = 0 \,, u(1,t) = 0 \tag{4.1.2}$$

$$u(x,0) = g(x), u_t(x,0) = 0 \tag{4.1.3}$$

Where $u(x) = g(x)$ is the equation of the initial position ***ACB***.

Equation (4.1.2) is the analytic statement that the endpoints **A** and **B** remain fixed. While equation (4.1.3) state that for $t = 0$ the string is in the initial position ***ACB***, and each particle starts its motion with an initial zero velocity.

4.2. Reduction to a One-dimensional Boundary Value Problem:

Let us try to find a solution of (4.1.1) in the form

$$u = \psi(x)\phi(t)$$

https://doi.org/ 10.2478/9788366675575-004

Substitution of this value of y in (4.1.1) gives

$$\psi(x)\frac{d^2\phi}{dt^2}=c^2\phi(t)\frac{d^2\psi}{dx^2}$$

Which can be put in the form

$$\frac{\frac{d^2\phi(t)}{dt^2}}{\phi(t)}=c^2\frac{\frac{d^2\psi(x)}{dx^2}}{\psi(x)}$$

The R.H.S is independent of t, and the L.H.S is independent of x.

Then either member is a constant which we designate as $-\lambda c^2$.

This gives us the two ordinary differential equation to solve

$$\frac{d^2\psi}{dx^2}+\lambda\psi=0,$$

$$\frac{d^2\phi}{dt^2}+\lambda c^2\phi=0$$

The initial conditions on u and ϕ are obtained from (4.1.2) and (4.1.3) and for ψ are as follows:

$$\psi(0)\phi(t)=0$$
$$\psi(1)\phi(t)=0$$

Whence $\psi(0)=0,\psi(1)=0$

https://doi.org/ 10.2478/9788366675575-004

Since $\phi(t) \neq 0$, We are thus led to the following one-dimensional boundary value problem: to determine a function $\psi(x)$ that will satisfy the differential equation:

$$\frac{d^2\psi}{dx^2} + \lambda\psi = 0 \tag{4.2.1}$$

and the boundary conditions

$$\psi(0) = 0, \psi(1) = 0$$ □

4.3. Solution of the Boundary Problem:

We see at once that $\psi \equiv 0$ is a solution. This gives the original problem of the vibrating string the trivial solution $u \equiv 0$, which means that the string remains at rest. Hereafter, we shall suggest a solution not identically zero when we refer to an explanation of our boundary problem.

To obtain the calculation, we have three cases to consider.

Case-I: $\lambda > 0$

From the elementary theory of differential equations, we know that the most general solution of (4.1.4) is

$$\psi(x) = A\cos\sqrt{\lambda}x + B\sin\sqrt{\lambda}x$$

The condition $\psi(0) = 0, \psi(1) = 0$ gives us $A = 0$ and either

https://doi.org/ 10.2478/9788366675575-004

(i) $B=0$ or **(ii)** $\sin\sqrt{\lambda}=0$

Now

(i) If $B=0$, we obtain the trivial solution $u\equiv 0$.

(ii) If $\sin\sqrt{\lambda}=0$, then $\lambda=n^2\pi^2$ (n, an integer) and the solution is

$$\psi(x)=B\sin n\pi x$$

Case-II $\lambda=0$,

The general solution of (4.1.4) is now $\psi(x)=Ax+B$.

But $\psi(0)=B=0$ and $\psi(1)=A=0$, so that we have the trivial solution again $\psi\equiv 0$.

Case-III: $\lambda<0$

The general solution of (4.1.4) is now

$$\psi(x)=Ae^{\sqrt{-\lambda}x}+Be^{-\sqrt{-\lambda}x}$$

Applying the initial conditions, we obtain

$\psi(0)=A+B=0$, Whence $B=-A$

$\psi(1)=A\left(e^{\sqrt{-\lambda}}-e^{-\sqrt{-\lambda}}\right)=0$, Whence $A=0$, $(\lambda\neq 0)$.

Therefore $A=B=0$ and we have the trivial solution again $\psi\equiv 0$.

Thus, we arrive at the following theorem,

https://doi.org/ 10.2478/9788366675575-004

Theorem-4.1

If $\lambda = n^2\pi^2$ (n, an integer), then the boundary problem

$$\frac{d^2\psi}{dx^2} + \lambda\psi = 0, \psi(0) = 0, \psi(1) = 0$$

has an infinitude of solutions

$\psi(x) = B\sin n\pi x$.

If $\lambda \neq n^2\pi^2$, then the only solution is the trivial one $\psi \equiv 0$. □

4.4. Construction of Green Function:

We now propose to show that every solution of our boundary problem (4.2.1) satisfies at the same time a linear integral equation. We observe first that the transformation method can not be used to determine an equivalent integral equation with the given boundary conditions.

In the present instance, to determine an equivalent integral equation, we first construct the Green function belonging to the boundary problem.

The given boundary problem for $\lambda = 0$ has only a trivial solution $\psi \equiv 0$. However, this is true only under the assumption tacitly made throughout, namely that ψ together with its first and second derivatives is continuous in the interval $[0,1]$.

Let us use the notation to denote this assumption. Drop now the assumption $\psi^{c''}$ and allow the derivative ψ' to be discontinuous at an arbitrary prescribed point ξ between 0 and 1, while ψ itself remains continuous.

Accordingly, we propose to determine a function u satisfying the following conditions

(a) ψ^{c} in $[0,1]$.

(b) $\psi^{c''}$ and $\dfrac{d^2\psi}{dx^2}=0$ in $[0,\xi]$.

$\psi^{c''}$ and $\dfrac{d^2\psi}{dx^2}=0$ in $[\xi,1]$.

(c) $\psi(0)=0, \psi(1)=0$.

The solution then must be of the form

$$\psi=\begin{cases}\alpha_0 x+\beta_0 & in\,[0,\xi]\\ \alpha_1 x+\beta_1 & in\,[\xi,1]\end{cases}$$

Similarly, let us use the notation C to denote all continuous functions and the notation C^n to represent all functions having continuous derivative up to the order n inclusive.

From **(c)**, we find

$$\psi(0)=\beta_0=0;\ \psi(1)=\alpha_1+\beta_1=0$$

Thus, the solution reduces to

$$\psi=\begin{cases}\alpha_0 x & in\ [0,\xi]\\ \beta_1(1-x) & in\ [\xi,1]\end{cases}$$

The condition **(a)** must be satisfied whence

$$\alpha_0\xi=\beta_1(1-\xi)$$

Therefore $\alpha_0=\rho(1-\xi)$ and $\beta_1=\rho\xi$.

The solution now takes the form

$$\psi=\begin{cases}\rho(1-\xi)x & in\ [0,\xi]\\ \rho\xi(1-x) & in\ [\xi,1]\end{cases}$$

Geometrically, the solution is represented by a broken line as in the adjoining figure.

https://doi.org/ 10.2478/9788366675575-004

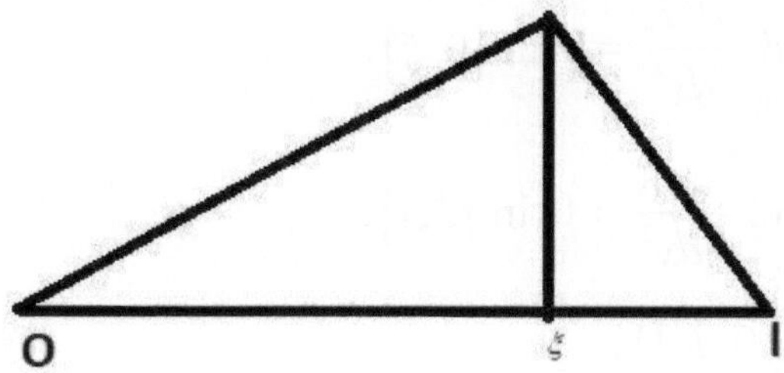

Fig.-5

For $x=\xi$, the derivative has a discontinuity measured by

$$\psi'(\xi-0)-\psi'(\xi+0)=\rho(1-\xi)+\rho\xi=\rho$$

We now impose the further condition that this discontinuity shall be $+1$.

Then $\rho=1$.

The function ψ so obtained is called Green's function for the boundary problem (4.2.1).

We use the notation $K(x,\xi)$ to represent this function.

We have thus the following:

Theorem-4.2

There exists one and only one function $K(x,\xi)$ which satisfies the conditions

(a) K^c in $[0,1]$.

https://doi.org/ 10.2478/9788366675575-004

(b) $K^{c''}$ and $\dfrac{d^2K}{dx^2}=0$ in $[0,\xi]$.

$K^{c''}$ and $\dfrac{d^2K}{dx^2}=0$ in $[\xi,1]$.

(c) $K(0)=0, K(1)=0$

(d) $[K']_{x=\xi+0}^{x=\xi-0}=+1$, $0<\xi<1$.

The formula gives this function

$$K(x,\xi)=\begin{cases}(1-\xi)x & for\ 0\le x\le\xi\\ \xi(1-x) & for\ \xi\le x\le 1\end{cases}$$

Let us use the notation

$$K_0(x,\xi)=(1-\xi)x$$
$$K_1(x,\xi)=\xi(1-x)$$

The properties **(a)** and **(d)** of $K(x,\xi)$may now be written as

$$K_0(\xi,\xi)=K_1(\xi,\xi)$$
$$K_0'(\xi,\xi)=-K_1'(\xi,\xi)=1$$

We can now prove the following:

Theorem-4.3

Green function is symmetric in x and ξ:

$$K(x,\xi)=K(\xi,x)$$

Proof:

Let $0 \le z_1 \le z_2 \le 1$

Then $K(z_1, z_2) = K_0(z_1, z_2) = (1 - z_2) z_1$

While $K(z_2, z_1) = K_1(z_1, z_2) = z_1(1 - z_2)$

Whence $K(z_1, z_2) = K(z_2, z_1)$ □

4.5. Equivalence between the Boundary value problem and Linear Integral Equations:

Take the equations

$$\frac{d^2\psi}{dx^2} = -\lambda\psi \, , \frac{d^2K}{dx^2} = 0$$

Multiply the first $-K$ and the second by ψ and add.

We obtain

$$\psi \frac{d^2K}{dx^2} - K\frac{d^2\psi}{dx^2} = \lambda\psi K \text{ , or}$$

$$\frac{d}{dx}\left(\psi\frac{dK}{dx} - K\frac{d\psi}{dx}\right) = \lambda\psi K$$

This equally holds in each of the two subintervals $[0, \xi]$ and $[\xi, 1]$. Integration over each of them gives

$$\left[\psi\frac{dK}{dx}-K\frac{d\psi}{dx}\right]_0^{\xi-0}=\lambda\int_0^{\xi-0}\psi K dx$$

$$\left[\psi\frac{dK}{dx}-K\frac{d\psi}{dx}\right]_{\xi+0}^{1}=\lambda\int_{\xi+0}^{1}\psi K dx$$

Both ψ and K vanish at 0 and 1.

$\psi,\frac{d\psi}{dx}$ and K are continuous functions of x over the whole interval $[0,1]$. Whence adding the last two equations, we obtain

$$\psi(\xi)\left[K'(\xi-0)-K'(\xi+0)\right]=\lambda\int_0^1 K(x,\xi)\psi(x)dx$$

But $K'(\xi-0)-K'(\xi+0)=1$ and so

$$\psi(\xi)=\lambda\int_0^1 K(x,\xi)\psi(x)dx$$

Now interchange x and ξ, and remember that

$K(x,\xi)=K(\xi,x)$.

Then

$$\psi(x)=\lambda\int_0^1 K(x,\xi)\psi(\xi)d\xi$$

This is a homogenous linear integral equation of the second kind for the determination of $\psi(x)$.

Every solution of the original boundary problem (4) satisfies this integral equation.

Hence, we have the followings.

Theorem-4.4

If $\psi(x)$ has continuous first and second derivatives and satisfies the boundary problem

$$\frac{d^2\psi}{dx^2}+\lambda\psi=0,\psi(0)=0,\psi(1)=0 \qquad (4.5.1)$$

Then $\psi(x)$ is continuous and satisfies the homogenous linear integral equation

$$\psi(x)=\lambda\int_0^1 K(x,\xi)\psi(\xi)d\xi$$

Where $K(x,\xi)$ is given by the formulas (4.5.1).

Let us now prove the following Converse.

Theorem-4.5

If $\psi(x)$ is continuous and satisfies the equation

https://doi.org/ 10.2478/9788366675575-004

$$\psi(x)=\lambda\int_0^1 K(x,\xi)\psi(\xi)d\xi \qquad (4.5.2)$$

Where $K(x,\xi)$ is given by the formula (4.5.1) Then u has continuous first and second derivatives and satisfies the boundary problem

$$\frac{d^2\psi}{dx^2}+\lambda\psi=0,\psi(0)=0,\psi(1)=0$$

Proof:

Let x and ξ range from 0 to 1 and $K(x,\xi)$ is discontinuous at $K(x,\xi)$.

Let us then write (4.5.2) in the form

$$\psi(x)=\lambda\int_0^x K_1(x,\xi)\psi(\xi)d\xi+\lambda\int_x^1 K_0(x,\xi)\psi(\xi)d\xi$$

Now we may apply the general rule for the differentiation of a definite integral concerning a parameter, Then

$$\frac{d\psi(x)}{dx}=\lambda\int_0^x K_1'(x,\xi)\psi(\xi)d\xi+\lambda\int_x^1 K_0'(x,\xi)\psi(\xi)d\xi$$

Since $K_0(x,x)=K_1(x,x)$

https://doi.org/ 10.2478/9788366675575-004

Moreover, since $K_1(x,\xi)$ and $K_0'(x,\xi)$ are continuous in their respective intervals, we see that $\frac{d\psi}{dx}$ is continuous.

A second differentiation gives

$$\frac{d^2\psi}{dx^2} = \lambda\int_0^x K_1''(x,\xi)\psi(\xi)d\xi + \lambda\int_x^1 K_0''(x,\xi)\psi(\xi)d\xi$$
$$+\lambda K_1'(x,x)\psi(x) - \lambda K_0'(x,x)\psi(x) = -\lambda\psi(x)$$

Since by our hypothesis on K ,we have

$K_1'' = K_0'' = 0$ and $K_1'(x,x) - K_0'(x,x) = -1$

Moreover, since $\psi(x)$ is continuous, it implies $\frac{d^2\psi}{dx^2}$ is also continuous.

We have further

$$\psi(0) = \lambda\int_0^1 K(0,\xi)\psi(\xi)d\xi = 0$$

Since $K(0,\xi) = 0$ and

$$\psi(1) = \lambda\int_0^1 K(1,\xi)\psi(\xi)d\xi = 0$$

Since $K(1,\xi) = 0$

https://doi.org/ 10.2478/9788366675575-004

We know the solution of the given BVP (4.5.1). This knowledge combined with **theorem-4.4** and **4.5** provides the following:

Theorem-4.6

Only when $\lambda = n^2\pi^2$ (n-an integer) does the integral equation

$$\psi(x) = \lambda\int_0^1 K(x,\xi)\psi(\xi)d\xi$$

have a solution not identically zero

$\psi(x) = B\sin n\pi x$.

Let's compare these results with the results obtained in the preceding chapter for the general homogenous integral equation. We see that the characteristic constants for this particular problem are $\lambda = n^2\pi^2 = \lambda_n$ and that they are of the index $q = 1$.

The following fundamental function belonging to λ_n is $\varphi(x) = \sin n\pi x$.

The kernel is a symmetric one, so that $K(x,\xi) = K(\xi,x)$.

https://doi.org/ 10.2478/9788366675575-004

And therefore, the associated equation functions are therefore

$\overline{\varphi}(x) = \sin n\pi x$.

4.6. Constrained Vibrations of an Elastic String:

4.6.1. The differential Equation of the problem:

Let us suppose that an exterior force $\mu H(x,t)$ acts on each particle of mass μ in the y-direction then it is known from the mathematical theory of vibrating string that the equations of motion of the string are

$$\frac{\partial^2 u}{\partial t^2} = c^2 \frac{\partial^2 u}{\partial x^2} + H(x,t) \tag{4.6.1}$$

$$u(0,t) = 0 \text{ , } u(1,t) = 0 \tag{4.6.2}$$

$$u(x,0) = g(x) \text{ , } u_t(x,0) = 0 \tag{4.6.3}$$

Let us suppose now that $H(x,t)$ is harmonic that is

$$H(x,t) = c^2 r(x)\cos(\beta t + \gamma), \ (c \neq 0)$$

Let us find, If possible, a solution of the form

$$u(x,t) = \psi(x)\cos(\beta t + \gamma)$$

Substitute this value of u in (4.6.1) and put $c^2\lambda = \beta^2$, we find

$$\frac{d^2 u}{dx^2} + \lambda\psi + r(x) = 0 \tag{4.6.4}$$

https://doi.org/ 10.2478/9788366675575-004

While from (4.6.2), we derive the boundary condition $\psi(0)=0\,,\psi(1)=0$ (4.6.5)

4.7. Equivalence Between Boundary value problem and Linear Integral Equations:

Construct as before the Green's function $K(x,\xi)$, Then

$$\frac{d^2k}{dx^2}=0\ ,\ \frac{d^2\psi}{dx^2}=-\lambda\psi-r$$

Multiply the first of these by u, the second by $-k$, and add,

We obtain $\psi\frac{d^2k}{dx^2}-k\frac{d^2\psi}{dx^2}=\lambda\psi k+rk$

Which may be written

$$\frac{d}{dx}\left(\psi\frac{dk}{dx}-k\frac{d\psi}{dx}\right)=\lambda\psi k+rk$$

Proceeding as before with the integration from 0 to ξ and from ξ to 1 , we find

$$\psi(\xi)=\lambda\int_0^1 K(x,\xi)\psi(x)dx+\int_0^1 K(x,\xi)r(x)dx$$

Interchange x and ξ, then on account of the symmetry of K, we have

https://doi.org/ 10.2478/9788366675575-004

$$\psi(x)=\lambda\int_0^1 K(x,\xi)\psi(\xi)d\xi+\int_0^1 K(x,\xi)r(\xi)d\xi$$

If, now, we put $\int_0^1 K(x,\xi)r(\xi)d\xi=f(x)$ (4.7.1)

We have

$$\psi(x)=f(x)+\lambda\int_0^1 K(x,\xi)\psi(\xi)d\xi \quad (4.7.2)$$

Which is a non-homogenous linear integral equation of the second kind.

It is satisfied by every solution of the boundary problem (4.6.4) and (4.6.5).

If we will proceed exactly as in the case of the homogenous equation. We can now show conversely that if u is continuous and satisfies the equation (4.7.2), where $f(x)$ is given by (4.7.1), then u has continuous first and second derivatives and satisfies the differential equation and the boundary conditions (4.6.5).

4.7.1. Remark on the solution of the BVP:

Equation (4.5.2) and (4.7.2) have the same kernel, namely the Green's function, which we have constrained knowing that the characteristic constants for (4.5.2), and hence for (4.7.2), are $\lambda_n = n^2\pi^2$. We obtain from the general theory the following results for (4.7.2):

Case-I

If $\lambda \neq n^2\pi^2$, then (4.7.2) has a unique solution.

Case-II:

If $\lambda = n^2\pi^2$, there is in general, no solution.

A solution exists only when the condition

$$\int_0^1 f(x)\sin n\pi x\, dx = 0 \tag{4.7.3}$$

is satisfied.

This condition is what

$$\int_a^b f(x)\bar{\phi}_\alpha(x)dx = 0$$

becomes for this special problem.

If (4.7.1) is satisfied, then (4.7.2) has a solution.

The BVP $\frac{d^2\psi}{dx^2} + \lambda\psi + r(x) = 0, \psi(0) = 0\,,\ \psi(1) = 0$

was shown to be equivalent to the non-homogenous integral equation (4.7.2), If

https://doi.org/ 10.2478/9788366675575-004

$$f(x)=\int_0^1 K(x,\xi)r(\xi)d\xi \tag{4.7.4}$$

This enables us from **cases-I and II** for the integral equation to state that when $\lambda \neq n^2\pi^2$ the BVP has a unique solution, and there is, in general, no solution for $\lambda = n^2\pi^2$, but when (4.7.3) is satisfied, there are ∞ solutions.

If now, in (4.7.3) we substitute for $f(x)$ its value as a given by (4.7.4), this condition becomes

$$\int_0^1\int_0^1 K(x,\xi)r(\xi)\sin n\pi x\, d\xi\, dx = 0$$

Interchange the order of integration and remember that $\sin n\pi x$ is a solution of the homogenous equation for $\lambda = \lambda_n$:

$$\sin n\pi x = \lambda_n \int_0^1 K(x,\xi)\sin n\pi\xi\, d\xi$$

Or $\sin n\pi\xi = \lambda_n \int_0^1 K(x,\xi)\sin n\pi x\, dx$

{Since $K(x,\xi)=K(\xi,x)$ },

Then the double integral becomes $\dfrac{1}{\lambda_n}\int_0^1 r(\xi)\sin n\pi\xi\, d\xi = 0$

That is so that the boundary problem (4.6.4) may have a solution. It is necessary and sufficient that $r(x)$ satisfies the equation

$$\int_0^1 r(x)\sin n\pi x\, dx = 0 . \qquad \square$$

https://doi.org/ 10.2478/9788366675575-005

Chapter-5

Hilbert-Schmidt Theory

Solution expressed in terms of a set of fundamental functions

5.1. Introduction

The Fredholm theory of a linear integral equation with a parameter λ:

$$y(x)=f(x)+\lambda\int_a^b K(x,t)\,y(t)\,dt \qquad (5.1.1)$$

has been developed under the assumptions

(a) $K(x,t)$ real.

(b) $K(x,t)$ Continuous.

(c) $K(x,t)\neq 0$ on $R: a\leq x\leq b \;\; a\leq t\leq b$.

For the Hilbert –Schmidt theory a fourth assumption is made that the kernel is symmetric.

(d) $K(x,t)=K(t,x)$ (5.1.2)

It is clear that the results of the Fredholm theory still hold.

But, besides several new results will follow from the additional condition **(d)**. Hilbert first obtained these results.

https://doi.org/ 10.2478/9788366675575-005

Fredholm obtained the solution of **(a)** by considering it as the Limit of a set of linear equations. He did not carry through the limiting process but guessed at the solution and then showed independently of the linear equations that the guess was correct.

Hilbert started from the finite system of linear equations, of which (5.1.1) may be considered as a limit and carried through the limiting process in details. Schmidt in 1906, obtained Hilbert's directly without using the limiting process.

Hilbert and Schmidt do not make use of Fredholm's results. We shall, however, avail ourselves of these results whenever they lead to simplifications of the proofs.

The first fundamental theorem of Hilbert's theory is the following:

Theorem-5.1

Every symmetric kernel has at least one characteristic constant (real or imaginary).

Proof:

Our proof of this theorem involves several lemmas, which we will give first. We provide an example due to

https://doi.org/ 10.2478/9788366675575-005

Kowalewski, which illustrates that a kernel that is not symmetric does not necessarily have a characteristic constant. The kernel

$$K(x,t) = \sin(\pi x)\cos(\pi t)\,,\ [0,1]$$

has no characteristic constant; if

$$y(x) = \lambda \int_0^1 \sin(\pi x)\cos(\pi t)\, y(t)\, dt \tag{5.1.3}$$

has no solution other than $u \equiv 0$.

An equation may be written

$$y(x) = \lambda \sin(\pi x) \int_0^1 \cos(\pi t)\, y(t)\, dt \equiv c \sin \pi x$$

Where c is a constant yet to be determined. Substitute this expression for $y(x)$ in (5.1.3), we obtain

$$c \sin \pi x = c\lambda \sin \pi x \int_0^1 \cos(\pi t)\sin(\pi t)\, dt$$

But $\int_0^1 \cos(\pi t)\sin(\pi t)\, dt = 0$.

Therefore $c \sin \pi x \equiv 0$.

Whence $c \equiv 0$.

and $y \equiv 0$.

https://doi.org/ 10.2478/9788366675575-005

Thus, this particular kernel has no characteristic constants. This shows that *theorem-5.1* indeed states a property peculiar to symmetric kernels. □

5.2. Iterations of symmetric kernels:

The iterated kernel $K_n(x,t)$ was defined by

$$K_i(x,t) = K(x,t)$$
$$K_n(x,t) = \int_a^b K(x,s)K_{n-1}(s,t)\,ds \qquad (5.2.1)$$

By successive applications of this recursion formula, we find

$$K_n(x,t)$$
$$= \int_a^b \dots \int_a^b K(x,s_1)K(s_1,s_2).....K(s_{n-1},t)\,ds_1 ds_2......ds_{n-1} \qquad (5.2.2)$$

As we have shown that

$$K_{n+p}(x,t) = \int_a^b K_n(x,s)K_p(s,t)\,ds \qquad (5.2.3)$$

We now establish the following lemmas:

Lemma-5.1

If $K(x,t)$ is symmetric, then $K_n(x,t)$ is symmetric.

Proof:

https://doi.org/ 10.2478/9788366675575-005

By (5.2.2), we have

$$K_n(x,t)=\int_a^b....\int_a^b K(t,s_1)K(s_1,s_2).....K(s_{n-1},x)ds_1ds_2......ds_{n-1}$$

$$=\int_a^b....\int_a^b K(s_1,t)K(s_2,s_1).....K(x,s_{n-1})ds_1ds_2......ds_{n-1}$$

On account of the symmetry of $K(x,t)$

$$=\int_a^b....\int_a^b K(x,s_{n-1})....K(s_2,s_1)K(s_1,t)ds_1ds_2......ds_{n-1}$$

Now make a change of notation.

For $s_1,s_2......,s_{n-2},s_{n-1}$

Put $s_{n-1},s_{n-2},.....,s_2,s_1$

Then

$$K_n(x,t)=\int_a^b....\int_a^b K(x,s_1)K(s_1,s_2).....K(s_{n-1},x)ds_1ds_2......ds_{n-1}$$ □

$$=K_n(x,t)$$

Lemma-5.2

If $K(x,t)$ satisfies the condition **(a), (b), (c), (d),** then $K_n(x,t)\neq 0$ on R.

Proof:

Suppose $K_n(x,t) \equiv 0$ on R, and Suppose that n is the lowest index for which an iterated kernel vanishes identically.

$$\left.\begin{array}{l} K(x,t) \neq 0 \\ K_2(x,t) \neq 0, \\ \ldots\ldots, \\ K_{n-1}(x,t) \neq 0, \\ K_n(x,t) \equiv 0 \end{array}\right. \quad \text{on } R.$$

Then indeed, by **(iii)**, we have $n > 1$.

Then from (5.2.3), it follows that all following iterated kernels are identically zero.

Then $K_{n+1}(x,t) \equiv 0$ on R .

Now either n or $n+1$ must be an even number $2m$:

$2m = n$ or $n+1$

Then $K_{2m}(x,t) \equiv 0$ on R.

But by (5.2.3),

$$K_{2m}(x,t) = \int_a^b K_m(x,s) K_m(s,t)\, ds$$

This equality holds when $t = x$,Then

https://doi.org/ 10.2478/9788366675575-005

$$K_{2m}(x,x) = \int_a^b K_m(x,s) K_m(s,x)\, ds$$

$$= \int_a^b \left[K_m(x,s)\right]^2 ds$$

On account of the symmetry of $K(x,t)$.

But $K_{2m}(x,t) \equiv 0$ on R.

Therefore, since $K_m(x,t)$ is a real function of x and t, so we have

$K_m(x,t) \equiv 0$ on R.

But $m = \frac{n}{2}$, or $\frac{n+1}{2}$

Hence $n - m = \frac{n}{2}$ or $\frac{n-1}{2} > 0$ for $n > 1$.

Therefore $m > n$.

Thus, we have contradicted the assumption that $K_n(x,t)$ was the first iterated function to vanish, and the Lemma is proved.

It is now clear that if $K(x,t)$ satisfies the condition **(a), (b),(c), (d),** then $K_n(x,t)$ satisfies the same conditions.

Corollary: $u_{2m} > 0$.

https://doi.org/ 10.2478/9788366675575-005

Proof: $u_{2m} = \int_a^b K_{2m}(x,x)\,dx$

Now $K_{2m}(x,x) > 0$,by (20), Since $K_m(x,t) \neq 0$. □

5.3. Orthogonality Theorem:

From the Fredholm theory

(i) Two associated kernels $\tilde{K}(x,t), K(x,t)$:

$$\tilde{K}(x,t) = K(t,x)$$

have the same characteristic constants with the same indices. If then

(ii) Let λ_0 and λ_1 $(\lambda_0 \neq \lambda_1)$ are two characteristic constants of $K(x,t)$ and hence of $K(t,x)$. $\phi_0(x)$ is a fundamental function of $K(x,t)$ for λ_0, and $\tilde{\phi}_1(x)$ is a fundamental function of $\tilde{K}(x,t)$ for λ_1,

it follows that

$$\int_a^b \phi_0(x)\tilde{\phi}_1(x)\,dx = 0$$

But if the kernel $K(x,t)$ is symmetric, then $\tilde{\phi}_1(x)$ is also a fundamental function of $K(x,t)$ for λ_1. If, then, we write $\tilde{\phi}_1 = \phi_1(x)$, we have

$$\int_a^b \phi_0(x)\phi_1(x)\,dx = 0$$

Definition:

Two functions $\phi_1(x)$, $\phi_2(x)$ continuous on the interval, $[a,b]$ are said to be *orthogonal* on $[a,b]$, If $(\phi_1\phi_2) = \int_a^b \phi_1(x)\phi_2(x)\,dx = 0$.

We have thus proved the following theorem which is called the *orthogonality theorem*.

Theorem-5.2

If $K(x,t)$ is symmetric and $\phi_0(x), \phi_1(x)$ are the fundamental function of $K(x,t)$ for λ_0 and λ_1 respectively $(\lambda_0 \neq \lambda_1)$. Then $\phi_0(x)$ and $\phi_1(x)$ are orthogonal on the interval $[a,b]$.

https://doi.org/ 10.2478/9788366675575-005

As an illustration of the **theorem-5.2**, take the symmetric kernel $K(x,t)$:

$$K(x,t)=\begin{cases}(1-t)x & ,[0,t]\\(1-x)t & ,[t,1]\end{cases}$$

Which has the characteristic constants

$\lambda_n = n^2\pi^2$, $n = 1,2,....$

and the corresponding fundamental function

$\phi_n(x)=\sin n\pi x$.

It is then well known from the theory of definite integrals that

$$\int_0^1 \sin n\pi x \sin p\pi x \, dx = 0,(n \neq p) \tag{5.3.1}$$

The Reality of Characteristic Constants:

Using the preceding theorem, we can now prove the following theorem due to Hilbert.

Theorem-5.3

If $K(x,t)$ is real and symmetric continuous and $\neq 0$, then all of the characteristic constants are real.

Proof:

Suppose that there is a characteristic constant λ_0 not real:

$\lambda_0 = \alpha_0 + i\beta_0$, $\beta_0 \neq 0$

Then the homogenous equation

$$y(x) = \lambda_0 \int_a^b K(x,t)\, y(t)\, dt$$

has at least one continuous solution $\phi(x) \equiv 0$:

$$\phi(x) = (\alpha_0 + i\beta_0) \int_a^b K(x,t)\, \phi(t)\, dt \tag{5.3.2}$$

(i) Suppose $\phi(x)$ is real, then separating the real and imaginary parts of (5.3.2)

We obtain

$$\phi(x) = \mu_0 \int_a^b K(x,t)\, \phi(t)\, dt \tag{5.3.3}$$

$$0 = \beta_0 \int_a^b K(x,t)\, \phi(t)\, dt \tag{5.3.4}$$

From (5.3.4)

$$\int_a^b K(x,t)\, \phi(t)\, dt = 0$$

Since $\beta_0 \neq 0$, Then from (5.3.3), $\phi(x) \equiv 0$

Which contradicts the assumption that $\phi(x) \neq 0$.

Therefore $\phi(x)$ cannot be real.

(ii) Suppose $\phi(x)=v_1(x)+iv_2(x)$, where $v_1(x)$ and $v_2(x)$ are real.

Then (5.3.2) becomes

$$v_1(x)+iv_2(x) = (\alpha_0+i\beta_0)\int_a^b K(x,t)\left[v_1(t)+iv_2(t)\right]dt \tag{5.3.5}$$

Separate (5.3.5) into its real and imaginary parts.

We obtain

$$v_1(x)=\alpha_0\int_a^b K(x,t)v_1(t)dt-\beta_0\int_a^b K(x,t)v_2(t)dt \tag{5.3.6}$$

$$v_2(x)=\alpha_0\int_a^b K(x,t)v_2(t)dt+\beta_0\int_a^b K(x,t)v_1(t)dt \tag{5.3.7}$$

Multiply both sides of (5.3.7) by $-i$ and add to (5.3.6), we obtain

$$v_1(x)-iv_2(x)=(\alpha_0-i\beta_0)\int_a^b K(x,t)\left[v_1(t)-iv_2(t)\right]dt$$

Therefore $v_1(x)-iv_2(x)\equiv\overline{\phi}(x)$ is also a fundamental function of $K(x,t)$ belonging to $\overline{\lambda}_0=\alpha_0-i\beta_0$. If we now apply the orthogonality **Theorem-5.3**, we obtain, since $\bar{\lambda}_0 \neq \lambda_0$,

$$\int_a^b \phi(x)\bar{\phi}(x)dx = 0 ,$$

Which may be rewritten

$$\int_a^b \left[v_1^2(x)+v_2^2(x)\right]dx = 0$$

But $v_1(x)$ and $v_2(x)$ are real functions.

Therefore

$$v_1^2(x)+v_2^2(x) = 0 ,$$

and hence

$$v_1(x) \equiv 0 , v_2(x) \equiv 0 .$$

and therefore $\phi(x) \equiv 0$.

Which constitutes a contradiction. Therefore λ_0 cannot be of the form $\lambda_0 = \alpha_0 + i\beta_0$ $(\beta_0 \neq 0)$ and thus λ_0 must be real.

This result might have been foreseen from the analogy with the finite system of linear equations:

$$y(t_i) - \lambda h \sum_{j=1}^{n} K(t_i, t_j) y(t_j) = f(t_i) , (i = 1, 2, .., n)$$

https://doi.org/ 10.2478/9788366675575-005

With determinants

$$\Delta = \begin{vmatrix} 1-\lambda hK_{11} & -\lambda hK_{12} & \cdots & -\lambda hK_{1n} \\ -\lambda hK_{21} & 1-\lambda hK_{22} & \cdots & -\lambda hK_{2n} \\ \vdots & \vdots & \cdots & \vdots \\ -\lambda hK_{n1} & -\lambda hK_{n2} & \cdots & 1-\lambda hK_{nn} \end{vmatrix} \qquad (5.3.8)$$

But it is well known that if K_{ij} is real and $K_{ij} = K_{ji}$ then all of the roots of (5.3.8) are real.

In the Fredholm theory, we proved that $q \le r$ the index r is the multiplicity of the root λ_0 of $D(\lambda_0)=0$. For a symmetric kernel, Hilbert has proved the following more definite theorem, which we state here without proof. □

Theorem-5.4

For a real symmetric kernel, the index q of a characteristic constant λ_0 is always equal to its multiplicity $r : q = r$.

5.4. An Existence theorem for Nonlinear Integral Equation of Fredholm type:

Consider the problem of establishing criteria for the existence of solutions for the nonlinear equation:

https://doi.org/ 10.2478/9788366675575-005

$$y(x) = f(x) + \lambda \int_a^b K[x, s, y(s)] ds \qquad (5.4.1)$$

Where λ is a parameter.

From the linear Volterra and Fredholm equations theory, we know that the parameter λ plays a significant role. Perhaps the essential difference between the two equations is found in the fact that, for bounded kernels, Integrable functions and a finite range of integrations. The solution in the Volterra case is an entire function of λ, while in the Fredholm case, it is a meromorphic function of the parameter with singularities at the zeros of the Fredholm determinant $D(\lambda)$. This same difference is observed in general between the two cases when the integral equation is nonlinear.

To establish criteria under which a solution exist for (5.4.1), we make assumption similar to those introduced for equations of Volterra type.

These are as follows:

(a) The function $f(x)$ is bounded in the interval: $a \le x \le b$ that is $|f(x)| < f$.

(b) The kernel $K(x, y, z)$ is integrable and bounded,

https://doi.org/ 10.2478/9788366675575-005

$$|K(x,y,z)| < k \tag{5.4.2}$$

In the domain $D: a \le x \le b, |z| < c$.

(c) $K(x,y,z)$ satisfies the Lipschitz condition in D, namely

$$|K(x,y,z) - K(x,y,z')| < M|z - z'| \tag{5.4.3}$$

By successive approximations, we now have

$$y_0(x) = f(x) - f(a)$$

$$y_1(x) = f(x) + \lambda \int_a^b K[x,s,y_0(s)]ds \tag{5.4.4}$$

and in general

$$y_n(x) = f(x) + \lambda \int_a^b K[x,s,y_{n-1}(s)]ds .$$

From these, we obtain

$$y_1 - y_0 = \lambda \int_a^b K[x,s,y_0(s)]ds + f(a)$$

$$y_2 - y_1 = \lambda \int_a^b \{K[x,s,y_1(s)] - K[x,s,y_0(s)]\} ds$$

$$\dots \quad \dots \quad \dots$$

$$y_n - y_{n-1} = \lambda \int_a^b \{K[x,s,y_{n-1}(s)] - K[x,s,y_{n-2}(s)]\} ds$$

From the conditions given above, we have

https://doi.org/ 10.2478/9788366675575-005

$$|y_1 - y_0| < |\lambda| K(b-a) + |f(a)|$$
$$= |\lambda|(b-a)K\left[1 + \frac{f}{|\lambda| K(b-a)}\right]$$
$$= |\lambda| m(b-a)$$

Where,

$$m = K\left|1 + \frac{f}{|\lambda| K(b-a)}\right|.$$

From this inequality and the Lipschitz condition, we get

$$|y_2 - y_1| < |\lambda| m \int_a^b |y_1 - y_0| ds < |\lambda|^2 M_m (b-a)^2 < |\lambda|^2 k^2 (b-a)^2$$

Where k is the larger of the two number M and m.

Similarly, we obtain the inequalities

$$|y_3 - y_2| < |\lambda|^3 k^3 (b-a)^3$$
$$|y_n - y_{n-1}| < |\lambda|^n k^n (b-a)^n$$

A majority for the series

$$y(x) = y_0 + [y_1(x) - y_0(x)] + .. \\ ... + [y_n(x) - y_{n-1}(x)] + ... \quad (5.4.5)$$

is furnished by the sum

https://doi.org/ 10.2478/9788366675575-005

$$y = f + \sum_{n=1}^{\infty} |\lambda|^n k^n (b-a)^n$$

and thus, the series converges uniformly for all values of λ for which we have

$$|\lambda| < \frac{1}{k(b-a)} \tag{5.4.6}$$

Although the condition (5.4.6) is equivalent to that obtained when equation (5.4.1) is linear, the role played by λ the case $f(x) \equiv 0$ is quite different in nonlinear equations from that which it has in the linear case. This difference will be shown in the next section.

5.5. A particular example:

As a particular case, we shall consider the equation

$$y(x) = \lambda \int_a^b (x-t) y^2(t) dt \tag{5.5.1}$$

Without loss of generality, we can set $\lambda = 1$, for if $z(x)$ is any solution of (5.5.1) then $y(x) = \lambda z(x)$ is a solution of the equation:

$$y(x) = \int_a^b (x-t) y^2(t) dt \tag{5.5.2}$$

Since it can be written

https://doi.org/ 10.2478/9788366675575-005

$$y(x)=x\int_0^1 y^2(t)\,dt-\int_0^1 t\,y^2(t)\,dt$$

It is clear that its solution is a linear function of x that is $y(x)=px+q$ (5.5.3)

where p and q are to be determined.

Introducing the function into (5.5.2), we obtain

$$px+q=x\left(\frac{1}{3}p^2+pq+q^2\right)-\left(\frac{1}{4}p^2+\frac{2}{3}pq+\frac{1}{2}q^2\right) \quad (5.5.4)$$

Equating the coefficient of x and constant terms of simplifying, we have the following system of equation for the determination of p and q:

$$p^2+3pq+3q^2=3p \quad (5.5.5)$$

$$3p^2+8pq+6q^2=-12q \quad (5.5.6)$$

If p and q are points in a Cartesian system of coordinates, both (5.5.5) and (5.5.6) represent ellipses that pass through the origin $p=q=0$.

These ellipses are graphically represented in **figure-1** $\left[c_1=(5),c_2=(6)\right]$ and are observed to intersect in two real points: the origin and the point $p=(-6,12)$.

https://doi.org/ 10.2478/9788366675575-005

Therefore equation (5.5.2) has one real solution other than the trivial one, $y(x)=0$ that is

$$y_1(x)=12x-6 \tag{5.5.7}$$

But it also has two complex solutions obtained from the intersections of the ellipses in the points: $(3+3i,-6)$ and $(3-3i,-6)$; that is

$$\begin{aligned} y_2(x) &= -6x+3(1+i) \\ y_3(x) &= -6x+3(1-i) \end{aligned} \tag{5.5.8}$$

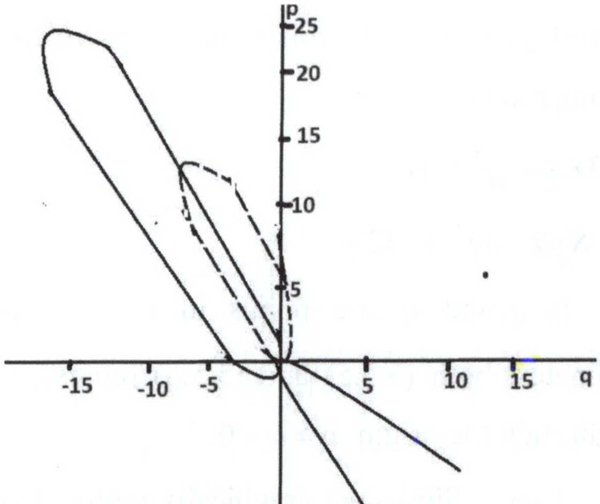

Figure-1

By a simple extension of this example, one can derive the following theorem:

The equation

$$y(x)=\int_a^b\big(A(t)x+B(t)\big)y^2(t)\,dt \tag{5.5.9}$$

In which $A(t)$ and $B(t)$ are real functions integrable in the interval (a,b) has three solutions other than $y(x)=0$ at least one of which is real.

By an argument similar to that used in the example, we obtain the following system of equations defining p and q in the solution:

$y(x)=px+q$

$$\begin{aligned} p^2\int_a^b t^2A(t)\,dt+2pq\int_a^b tA(t)\,dt+q^2\int_a^b A(t)\,dt&=p\\ p^2\int_a^b t^2B(t)\,dt+2pq\int_a^b tB(t)\,dt+q^2\int_a^b B(t)\,dt&=q \end{aligned} \tag{5.5.10}$$

If this system is solved for $p(or\ q)$, the resulting equation that determines p is a quartic, one root of which is that when the factor reduces the quartic p, the resulting equation is a cubic with real coefficients and must have one real root. The theorem results from this observation. □

As a second example, consider the equation

$$y(x)=\int_{0}^{1}(x+t)y^2(t)\,dt \qquad (5.5.11)$$

The first equation in (5.5.10) reduces to (5.5.5), but the second becomes

$$3p^2+8pq+6q^2=12q$$

Which is the ellipse C_3 shown in **figure-1.**

We see from the graphs that there is only one real solution corresponding to the point Q. This solution to three decimal approximation is the following

$$y(x)=0.726x+0.461//$$

The examples of this section illustrate the difference between linear and nonlinear integral equations. Thus, if $K(x,t)$ is a symmetric function, the solution of the equation:

$$y(x)=\lambda\int_{a}^{b}K(x,t)y(t)\,dt \qquad (5.5.12)$$

will depend upon the parameter λ and will all be real.

If $K(x,t)$ is a skew-symmetric function, the solution is all complex. In the examples given, however, one real solution and two complex solutions exist for both (5.5.11) and (5.5.12)

The fact that the kernel in the first case is skew-symmetric and in the second is symmetric had nothing to do with the situation.

Moreover, in the nonlinear equations, the solution was not dependent upon the parameter values of the parameter λ, a very conspicuous aspect of linear integral equations.

It should be pointed out that the method just given is also applicable to the solution of the equation:

$$y(x) = Ax + B + \lambda\int_a^b \left[A(t)x + B(t)\right] y^2(t)\,dt$$

Since the form of the solution remains unchanged, but in this case, the solution depends upon λ, although not in the same critical sense as in the theory of linear integral equations.

However, we can no longer say that there is a real solution since the system equations (5.5.10) are each augmented by an arbitrary constant.

In this case, there will be, in general, four non-trivial solutions.

The equation $y(x) = \lambda\int_a^b K(x,t)\, y^n(t)\,dt$ **:**

The particular equation solved in the preceding section is a special case of the following more general one;

$$y(x)=\lambda\int_{a}^{b}K(x,t)y^{n}(t)dt \tag{5.5.12}$$

The solution of which we shall now consider except in the linear case when $n=1$, the parameter λ can be set equal to unity without loss of generality.

For if we make the transformation

$$y(x)=\lambda^{m}z(x)$$

Equation (5.5.12) reduces to the following

$$\lambda^{m}z(x)=\lambda^{mn+1}\int_{a}^{b}K(x,t)z^{n}(t)dt \tag{5.5.13}$$

If m is chosen so that $mn+1=m$, that is

$$m=-\frac{1}{(n-1)},\quad n\neq 1$$

We get

$$z(x)=\int_{a}^{b}K(x,t)z^{n}(t)dt \tag{5.5.14}$$

We shall assume that the kernel has the following form:

$$K(x,t)=\sum_{i=1}^{R}X_{i}(x)Y_{i}(t) \tag{5.5.15}$$

https://doi.org/ 10.2478/9788366675575-005

When the functions: $X_1(x), X_2(x), \ldots$ are considered to form a linearly independent set. Introducing this kernel in to (5.5.12) and setting $\lambda = 1$, we get

$$\begin{aligned} y(x) &= \sum_{i=1}^{R} X_i(x) \int_a^b Y_i(t) y^n(t)\,dt \\ &= \sum_{i=1}^{R} X_i(x) p_i \end{aligned} \quad (5.5.16)$$

Where the p_i's are constants defined by

$$p_i = \int_a^b Y_i(t) y^n(t)\,dt \quad (5.5.17)$$

When (5.5.16) is introduced in to (5.5.12), we have

$$\sum_{i=1}^{R} X_i(x) p_i = \sum_{i=1}^{R} X_i(x) \int_a^b Y_i(t) \left[\sum_{j=1}^{R} X_j(t) p_j \right]^n dt$$

Since $X_i(x)$ are by assumption linearly independent functions of x, we can their coefficients and thus obtain the following system of equations for the determination of the constants p_i;

$$p_i = \int_a^b Y_i(t) \left[\sum_{j=1}^{R} X_j(t) p_j \right]^n dt \,,\ i = 1, 2, \ldots R \quad (5.5.18)$$

We thus have a set of R algebraic equations of nth degree for the determination of the constants p_i.

These equations will have, in general, R^n solutions, one of which is the trivial case $p_i = 0$, since these solutions are obtained from R algebraic equations with real coefficients in each of the variables and each of the degree $N = R^n$.

Since the form of these equations is as follows:

$$p_i \left(A_i p_i^{N-1} + B_i p_i^{N-2} + ...\right) = 0 \qquad (5.5.19)$$

At least one real set of values of the p_i provided $N-1 = R^n - 1$ is an odd integer, that is to say. If R is an even integer.

We thus have the following theorem:

If in equation (5.5.12), the kernel has the form:

$$K(x,t) = \sum_{i=1}^{R} X_i(x) Y_i(t)$$

The function $X_i(x)$ is linearly independent. There will exist in general $N-1 = R^n - 1$ solutions other than the trivial one $u = 0$, and of these, at least one will be a real solution provided R is an even integer.

When $n = 2$, equation (5.5.18) reduces to the form

https://doi.org/ 10.2478/9788366675575-005

$$p_i = F(p,p), i = 1,2,...R \tag{5.5.20}$$

Where $F(p,p)$ is the real quadratic form:

$$F(p,p) = \sum_{jk} a_{ik}^{(i)} p_j p_k \tag{5.5.21}$$

In which

$$a_{ik}^{(i)} = a_{mj}^{(i)} = \int_a^b X_j(t) X_k(t) Y_i(t) dt \tag{5.5.22}$$

In this case there will exist $R^2 - 1$ solution with at least one real solution when R is an even integer. the actual determination of the solution is probably most easily achieved by forming the eliminates for each p_i.

(Using *Sylvester's dialytic method*)

Thus p_1 is eliminated between $i = 2$ and $i = 3$, and so on.

In this way, $n-1$ equations are obtained containing the remaining variables.

From these p_2 in turn is eliminated and the process is continued until the final eliminant is obtained, which will turn out to be an equation of degree $N = R^n$ in p_R of form (5.5.19) above.

This process is illustrated readily by considering equation (5.5.16) and (5.5.17), namely

$$\begin{aligned} &p^2+3pq+3q^2-3p=0 \\ &3p^2+8pq+6q^2+12q=0 \end{aligned} \qquad (5.5.23)$$

Multiplying each in turn by p, we obtain the following system of four equations in the variables p p^2 and p^3:

$$\begin{aligned} &p^2+(3q-3)p+3q^2=0 \\ &p^3+(3q-3)p^2+3q^2p=0 \\ &3p^2+8qp+6q^2+12q=0 \\ &3p^3+8qp^2+\left(6q^2+12q\right)p=0 \end{aligned} \qquad (5.5.24)$$

If these equations in p, p^2 and p^3 are to be consistent. The determinant must be set equals to zero, and we thus obtain the eliminant as follows:

$$\begin{vmatrix} 0 & 1 & (3q-3) & 3q^2 \\ 1 & (3q-3) & 3q^2 & 0 \\ 0 & 3 & 8q & \left(6q^2+12q\right) \\ 3 & 8q & \left(6q^2+12q\right) & 0 \end{vmatrix}=0 \qquad (5.5.25)$$

This reduces to the equation:

$$3q\left(3q^3-18q+108\right)=0$$

A similar computation for the system corresponding to equation (5.5.11) of section (5.5.16), namely

$$\begin{aligned} &p^2+3pq+3q^2-3p=0 \\ &3p^2+8qp+6q^2-12q=0 \end{aligned} \qquad (5.5.26)$$

This leads to the eliminant

$$3q\left(q^3+24q^2+222q-108\right)=0 \qquad (5.5.27)$$

5.6. The Equation of Bratu:

The equation of Bratu is the following nonlinear integral equation:

$$y(x)=\lambda\int_a^b G(x,t)e^{y(t)}dt \qquad (5.6.1)$$

Where the kernels are Green's function defined as follows:

$$\begin{aligned} G(x,t)&=\frac{(b-x)(t-a)}{b-a},t\le x \\ &=\frac{(b-t)(x-a)}{b-a},x\le t \end{aligned} \qquad (5.6.2)$$

One can show that the function $y(x)$ which satisfies this equation is any solution of the differential equation

$$\frac{d^2y}{dx^2}+\lambda e^y=0 \qquad (5.6.3)$$

This also meets the two-point boundary condition:

$$y(a)=y(b)=0$$

https://doi.org/ 10.2478/9788366675575-005

It will be convenient to assume that $a=0$ and that $\lambda>0$.

We shall now prove the following theorem (***Bratu***) For every value of λ taken between 0 and λ_1, where

$$\lambda_1=\frac{(1.8745....)^2}{b^2} \qquad (5.6.4)$$

Equation (5.6.1) has two real and distinct solutions. These curves C_1 and C_2 are of the parabolic form, concave to the x-axis and pass through the points $x=0$ and $x=b$.

As $\lambda\to\lambda_1$, the two curves tend towards a limiting curve C between them. When $\lambda=\lambda_1$, the limit curve is the unique solution of equation (5.6.1).

For $\lambda>\lambda_1$, the integral equation has no real solution.

Equation (5.6.3) can be integrated, and we have as a first integral

$$y'^2-m^2+2\lambda\left(e^y-1\right)=0 \qquad (5.6.5)$$

Where $m=y'(0)$.

Introducing the variable $t=1+\frac{m^2}{2\lambda}$

We get $y'^2=2\lambda\left(t-e^y\right)$ and thus the solution of (5.6.3) in the following form:

https://doi.org/ 10.2478/9788366675575-005

$$x = -\frac{1}{\sqrt{2\lambda}} \int_0^y \frac{dy}{\sqrt{t - e^y}} \tag{5.6.6}$$

The curve $y = y(x)$ increases to a maximum at $y = \log t$ and hence at this point

$$x = x_1 = \frac{1}{\sqrt{2\lambda}} \int_0^{\log t} \frac{dy}{\sqrt{t - e^y}} = \frac{1}{\sqrt{2\lambda}} \int_1^t \frac{du}{u\sqrt{t - u}} \ . \tag{5.6.7}$$

When $x > x_1$, y diminishes and since the curve is symmetric concerning the line $x = x_1$, y is zero when

$$x = 2x_1 = b = \sqrt{\frac{8}{\lambda t}} \log\left[\sqrt{t} + \sqrt{t-1}\right] \tag{5.6.8}$$

Since $m = 0, b = 0$ for $t = 1$ and $m = \infty, b = 0$ for $t = \infty$, b passes through a maximum value as m varies between 0 and ∞ .

To find this value, we compute:

$$\frac{db}{dt} = \sqrt{\frac{2}{\lambda t^2}} \left[\frac{1}{\sqrt{t-1}} - \frac{\log\left[\sqrt{t} + \sqrt{t-1}\right]}{\sqrt{t}} \right]$$

Equating $\frac{db}{dt}$ to zero, we now solve the resulting transcendental equation and thus find $t = 3.2766...$.

https://doi.org/ 10.2478/9788366675575-005

Denoting the maximum value of b by β.we now substitute t in (5.6.8) and thus obtain $\beta = \frac{h}{\sqrt{\lambda}}$, $h = 1.8745.....$

The theorem follows readily from these results. If we assumed a value of b and regarded λ it as a variable, there will be a value for each value t.

If $b = \beta$, then the initial slope of $y(x)$ given by Equation (5.6.5), denoted by μ and is found to be

$\mu = 2.1338....\sqrt{\lambda}$

For this value, there exists one curve, namely C, which passes through $x = 0$ and $x = \beta$.

But if $b < \beta$, that is to say, if $\lambda_1 > \lambda = \frac{h^2}{\beta^2}$, then there will be two curves, C_1 and C_2 which pass through $x = 0$ and $x = b$. For the first one, the slope m_1 will be greater than μ , and the slope m_2 will be less for the second.

If $\lambda_1 < \lambda$, then $b > \beta$, but since β is the maximum value which b can have, it is clear that there will be no solution in this case.

https://doi.org/ 10.2478/9788366675575-005

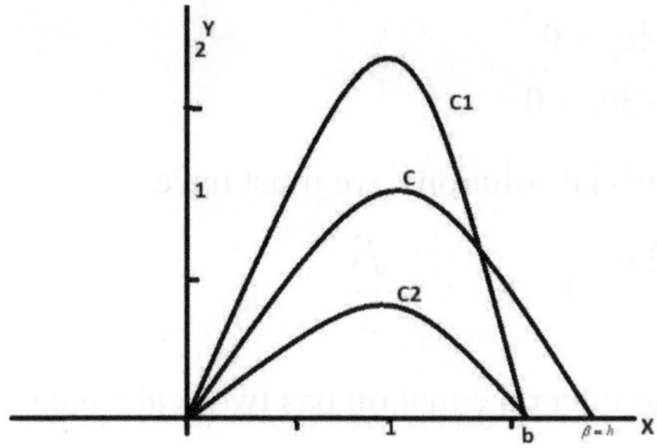

Figure-2

The situation is shown in **figure-2** where $\lambda = 1$ and $\lambda_1 = 1.275$.

Example-1

Find the Eigenvalues and Eigenfunctions of the integral equation with the symmetric kernel.

$$y(x) = \lambda \int_{-1}^{1} (x+t)\, y(t)\, dt$$

Solution:

Here $y(x) = \lambda c_1 x + \lambda c_2$, where

$$c_1 = \int_{-1}^{1} y(t)\, dt = 2\lambda c_2$$

$$c_2 = \int_{-1}^{1} t y(t)\, dt = \frac{2\lambda}{3} c_1$$

$$\therefore c_1 - 2\lambda c_2 = 0$$
$$2\lambda c_1 - 3c_2 = 0$$

For nontrivial solutions, we must have

$$\begin{vmatrix} 1 & -2\lambda \\ 2\lambda & -3 \end{vmatrix} = 0 \Rightarrow \lambda = \pm\frac{\sqrt{3}}{2}.$$

Thus, the integral equation has two real Eigen values

$\lambda = \frac{\sqrt{3}}{2}$ and $\lambda = -\frac{\sqrt{3}}{2}$.

If $\lambda = \frac{\sqrt{3}}{2}$,then $c_1 = \sqrt{3}c_2$ .the corresponding Eigen function is

$y_1(x) = 1 + \sqrt{3}x$.

Similarly, if $\lambda = -\frac{\sqrt{3}}{2}$,then $c_1 = -\sqrt{3}c_2$ and the corresponding

Eigen function is

$y_2(x) = 1 - \sqrt{3}x$.□

It can be seen that the functions $y_1(x)$ and $y_2(x)$ are orthogonal in $(-1,1)$.

Example-2

$$y(x) = \lambda \int_{-1}^{1} (x-t) y(t) dt$$

https://doi.org/ 10.2478/9788366675575-005

Solution:

Here

$$y(x) = \lambda c_1 x - \lambda c_2$$

Where ,

$$c_1 = \int_{-1}^{1} y(t)\,dt = -2\lambda c_2$$

$$c_2 = \int_{-1}^{1} t y(t)\,dt = \frac{2\lambda}{3} c_1$$

$$\therefore c_1 + 2\lambda c_2 = 0$$
$$2\lambda c_1 - 3c_2 = 0$$

The Eigen values are given by

$$\begin{vmatrix} 1 & 2\lambda \\ 2\lambda & -3 \end{vmatrix} = 0 \Rightarrow \lambda = \pm\frac{\sqrt{3}}{2} i$$

It implies that the Eigen values are purely imaginary.

Hence the Eigen functions are

$y_1(x) = 1 - \sqrt{3}ix$ and $y_2(x) = 1 + \sqrt{3}ix$. □

https://doi.org/ 10.2478/9788366675575-005

Exercises

Reduce the following equations to the equations of the second kind

1. $\int_0^x 3^{x-t} y(t)\,dt = x$

Ans.

$$y(x) = 1 - \ln 3 \int_0^1 3^{x-t} y(t)\,dt$$

2. $\int_0^x a^{x-t} y(t)\,dt = f(x), f(0) = 0$

Ans.

$$y(x) = f'(x) - \ln \mathrm{a} \int_0^x a^{x-t} y(t)\,dt$$

3. $\int_0^x \left(1 - x^2 + t^2\right) y(t)\,dt = \frac{x^2}{2}$

Ans.

$$y(x) = x + 2\int_0^x xy(t)\,dt$$

4. $\int_0^x \left(2 + x^2 - t^2\right) y(t)\,dt = x^2$

Ans.

https://doi.org/ 10.2478/9788366675575-005

$$y(x)=2x-2\int_0^x xy(t)\,dt$$

The First Bratu-Type Problem

5. Consider a Bratu-Type problem of this form

$$\frac{d^2y}{dx^2}-\pi^2e^y=0\ ,0<x<1$$

$$y(0)=y(1)=0$$

The Second Bratu-Type Problem

6. Consider a Bratu-Type problem of this form

$$\frac{d^2y}{dx^2}+\pi^2e^{-y}=0\ ,0<x<1\ ,y(0)=y(1)=0$$

The Third Bratu-Type Problem

7. Consider a Bratu-Type problem of this form

$$\frac{d^2y}{dx^2}-e^y=0\ ,0<x<1\ ,\ y(0)=y(1)=0$$

https://doi.org/ 10.2478/9788366675575-006

Chapter-6

The Fredholm Alternatives

In this chapter, we take a new look at the iterative method introduced earlier. Our goal will be to introduce the Neumann series. The terms of which involves Iterations of the kernel of an integral equation and use this series to derive the solution of the equation.

Throughout the chapter, we shall be concerned solely with the non-homogenous Fredholm equation

$$y(x)=f(x)+\lambda\int_a^b K(x,t)y(t)dt \quad , \ x\in[a,b]$$

Where K and f are real-valued continuous functions and λ is a constant. The kernel $K=K(x,t)$ need not be symmetric.

6.1. An example of Picard's method:

We first suggest that the reader review Picard's method to solve the non-homogenous Fredholm equation.

We now introduce a simple example to which we shall return at various points in this chapter.

https://doi.org/ 10.2478/9788366675575-006

It is used here to continue reviewing Picard's method and will also be used to exemplify the new procedure, which will appear later.

Example-1

Use Picard's method to find the solution of the equation

$$y(x)=1+\lambda\int_a^b x.t\,y(t)\,dt\ ,\ \ x\in[0,1]$$

The method provided the successive Iteration

$$y_0(x)=1$$

$$y_1(x)=1+\frac{\lambda x}{2}$$

$$\cdots\ \cdots\ \ \cdots$$

$$y_n(x)=1+\frac{\lambda x}{2}\left(1+\frac{\lambda}{3}+.....+\left(\frac{\lambda}{3}\right)^{n-1}\right)$$

As the reader will readily verify.

The kernel $K(x,t)=x.t$ is bounded by 1 on $[0,1]$.So y_n tends to the solution of the equation if $|\lambda|<1$.

Note:

In the above example:

https://doi.org/ 10.2478/9788366675575-006

(a) The sequence $\{y_n\}$ converges when the geometric series

$\sum_{n=0}^{\infty}\left(\frac{\lambda}{3}\right)^n$ Converges; that is, when $|\lambda|<3$.

The solution is then given by

$$y(x)=1+\frac{\lambda x}{2}\left(1-\frac{\lambda}{3}\right)^{-1}=1+\frac{3\lambda x}{2(3-\lambda)} \quad \text{for } x\in[0,1] \quad (6.1.1)$$

(b) The solution (6.1.1), and for the broadest possible range of λ, can be found most easily by using Fredholm Alternative method. For simple degenerate kernels, That method is the one advised.

The methods of this chapter come into their own in more complicated situations. □

Example-2

Show that if λ_1 is the eigenvalue of

$$y(x)=\lambda\int_a^b K(x,t)y(t)dt$$

of least modulus, K is symmetric and $b>a$.

Then $\lambda_1 \geq \left(\int_a^b\int_a^b K^2\right)^{-\frac{1}{2}} \geq [L(b-a)]^{-1}$.

https://doi.org/ 10.2478/9788366675575-006

Where b is the least upper bound of $|K|$ on $[a,b]$.

Show that these inequalities hold with when $a=0, b=1$ and either $K(x,t)=x.t$ or $K(x,t)=x^2t^2$ or $K(x,t)=x+t$.

Note: when the inequalities are strict.

6.2. Powers of an Integral Operator:

The material of this section only provides an alternatives way of looking at Picard's method.

The integral operator $k=k[a,b]$ associates with every continuous $F:[a,b]\to R$,

The function kF defined by

$$kF(x)=\int_a^b K(x,t)F(t)dt \text{ , } x\in[a,b]$$

Power of k is defined in an obvious way, using

$$k'=k$$

$$k^nF=k\left(k^{n-1}F\right) \text{ , } n\geq 2$$

(Of course, every k^nF must be continuous)

In particular

$$k^3F(x)=\int_a^b K(x,v)\int_a^b K(v,u)\int_a^b K(u,t)F(t)\,dt\,du\,dv$$

The Picard iteration may then be written

$$y_0(x)=f(x)$$
$$y_n(x)=f(x)+\lambda k y_{n-1}(x)\text{ , }(n\geq 1)$$

By induction, the reader is asked to provide,

$$y_n(x)=f(x)+\sum_{i=1}^{n}\lambda^i k^i f(x) \qquad (6.2.1)$$

So that, $y_n(x)-y_{n-1}(x)=\lambda^n k^n f(x)$, $(n\geq 1)$

It is straight forward to show directly that

$$\left|k^n f(x)\right|\leq L^n M(b-a)^n,(n\geq 1)$$

Where L and M bound on $|k|$ and $|f|$ respectively and the proof that $\{y_n\}$ converges uniformly to the solution of (N) may then be completed as previously.

The solution may thus be expressed

$$y(x)=f(x)+\sum_{n=1}^{\infty}\lambda^n k^n f(x)\text{ , }\left(x\in[a,b]\right) \qquad (6.2.2)$$

Which is certainly valid for $|\lambda|<\left[L(b-a)\right]^{-1}$.□

https://doi.org/ 10.2478/9788366675575-006

Example-3

If $K(x,t)=xt$ and $f(x)=1$, For each x and t in $[0,1]$, Calculate $k^n f(x)$ directly when $k=k[0,1]$.

6.3. Iterated Kernels:

Instead of iterating an operator, one can iterate under the integral sign by iterating the kernel.

Define, for x,t in $[a,b]$.

$$K_1(x,t)=K(x,t)$$

$$K_n(x,t)=\int_a^b K(x,u)K_{n-1}(u,t)\,du$$

We show by induction that

$$k^n F(x)=\int_a^b K_n(x,t)F(t)\,dt\ ,(n\geq 1) \qquad (6.3.1)$$

For any continuous $F:[a,b]\to R$.

The definition of k gives (6.3.1) for $n=1$ .So, suppose that equality holds for $n-1$.

Then, using Fubini's theorem to interchange the order of integration,

https://doi.org/ 10.2478/9788366675575-006

$$k^n F(x) = k\left(k^{n-1}F(x)\right)$$

$$= k\left(\int_a^b K_{n-1}(x,t)F(t)\,dt\right)$$

$$= \int_a^b K(x,u)\left(\int_a^b K_{n-1}(u,t)F(t)\,dt\right)du$$

$$= \int_a^b\left(\int_a^b K(x,u)K_{n-1}(u,t)\,du\right)F(t)\,dt$$

And the formula (6.3.1) is thus generally established.

Example: For $y(x) = 1 + \lambda\int_a^b x.t\, y(t)\,dt, x \in [0,1]$

Where $K(x,t) = x.t$ and integration is over $[0,1]$, show that

$K_n(x,t) = \frac{x^t}{3^{n-1}}$ for each x and t in $[0,1]$. □

6.4. Neumann series:

With the iterated kernels of the above section and equation (6.3.1) established, we are in a position immediately to rewrite the solution of (N) given as (6.2.2) in the following form:

$$y(x) = f(x) + \sum_{n=1}^{\infty}\lambda^n\int_a^b K_n(x,t)f(t)\,dt,\ \left(x \in [a,b]\right)$$

https://doi.org/ 10.2478/9788366675575-006

The convergence of the series being uniform for $|\lambda|<[L(b-a)]^{-1}$ where (as before) L is an upper bound on $|K|$ over $[a,b]$.

We may interchange summation and integration to give

$$y(x)=f(x)+\lambda\int_a^b N(x,t;\lambda)f(t)dt \qquad (6.4.1)$$

For $(x\in[a,b])$, Where

$$N(x,t;\lambda)=\sum_{n=1}^{\infty}\lambda^n K_{n+1}(x,t) \qquad (6.4.2)$$

For $(x,t\in[a,b])$ and $|\lambda|<[L(b-a)]^{-1}$.

The series appearing in (6.4.2), and also the series representation of y in (6.4.1) (when the substitution is made for $N(x,t;\lambda)$ from (6.4.2)) are called *Neumann series*.

The function $N(x,t;\lambda)$ is often termed the reciprocal or resolvent kernel as the solution of(N), given as (6.4.1) may be written as the reciprocal integral equation

$$f(x)=y(x)-\int_a^b N(x,t;\lambda)f(t)dt \,.\square$$

https://doi.org/ 10.2478/9788366675575-006

Example-4

In Example-1, Show that the kernel gives rise to the resolvent kernel

$$N(x,t;\lambda)=3xt(3-\lambda)^{-1},$$

When $|\lambda|<3$ and verify that produces the solution (6.1.1).

Example-5

Apply the methods of Picard, integral operator powers and Neumann series in turn to find the solution to

$$y(x)=1+\lambda\int_0^1(1-3xt)y(t)dt\ ,\ (x\in[0,1])$$

Example-6

Establish the identities

$$N(x,t;\lambda)=k(x,t)+\lambda\int_a^b k(x,u)N(u,t;\lambda)du$$

$$N(x,t;\lambda)=k(x,t)+\lambda\int_a^b k(u,t)N(x,u;\lambda)du$$

Where $x,t\in[a,b]$ and $|\lambda|<[L(b-a)]^{-1}$.

https://doi.org/ 10.2478/9788366675575-006

6.5. A Remark on the convergence of the Iterative method:

When discussing an iterative procedure, we have sufficient condition for the iteration to converse in an upper bound $|\lambda|$.

The curious reader will have wondered about the best upper bound, about a necessary condition. The answer is a simple one: the series provided by iterative methods in this set of notes converges when and only when $|\lambda| < |\lambda_1|$ the corresponding homogenous equation of least modulus is the eigenvalue of the corresponding homogenous equation.

Example-7

Show that the equation (N) has a unique solution expressible as a power series in λ. when its kernel $K = K(x,t)$ is degenerate and $|\lambda| < |\lambda_1|$ where λ_1 is the eigenvalue of the corresponding homogenous equation of least modulus.

6.6. Differentiation of function under an integral sign

Consider the function $I_n(x)$ defined by the relation

https://doi.org/ 10.2478/9788366675575-006

$$I_n(x)=\int_a^x (x-\eta)^{n-1} f(\eta)\,d\eta \tag{6.6.1}$$

Where η is a positive integer and a is a constant.

We know that

$$\frac{d}{dx}\int_{P(x)}^{Q(x)} F(x,\eta)\,d\eta=\int_{P(x)}^{Q(x)}\frac{\partial}{\partial x}\{F(x,\eta)\}\,d\eta$$
$$+F\left[x,Q(x)\right]\frac{dQ}{dx}-F\left[x,P(x)\right]\frac{dP}{dx}$$

Which is valid if F and $\frac{\partial F}{\partial x}$ are continuous functions of both x,η and the first derivative of $P(x)$ and $Q(x)$ are continuous.

Differentiation (6.6.1) under the integral sign, we have

$$\frac{dI_n}{dx}=(n-1)\int_a^x (x-\eta)^{n-2} f(\eta)\,d\eta\times\left[(x-\eta)^{n-2} f(\eta)\right]_{\eta=x} \tag{6.6.2}$$

$$\frac{dI_n}{dx}=(n-1)I_n \quad ,n>1$$

From the relation (6.6.1), we have

$$I_1(x)=\int_a^x f(\eta)\,d\eta\Rightarrow\frac{dI_1}{dx}=f(x) \tag{6.6.3}$$

Differentiation (6.6.2) successively m-times, we have

$$\frac{d^m I_n}{dx^m}=(n-1)(n-3).......(n-m)I_{n-m}\,,\; n>m$$

In particular, we have

$$\frac{d^{n-1}I_n}{dx^{n-1}} = (n-1)!\,I_1(x)$$

$$\Rightarrow \frac{d}{dx}\left(\frac{d^{n-1}I_n}{dx^{n-1}}\right) = (n-1)!\frac{dI_1}{dx} = (n-1)!\,f(x)$$

Thus, we have $I_1(x) = \int_a^x f(x_1)\,dx_1$

$$\frac{dI_2}{dx} = I_1 = \int_a^x f(x_1)\,dx_1$$

$$\Rightarrow I_2(x) = \int_a^x \int_a^{x_2} f(x_1)\,dx_1 dx_2 \tag{6.6.4}$$

In general, we have

$$I_n(x) = (n-1)!\int_a^x \int_a^{x_n} \ldots\ldots \int_a^{x_3} \int_a^{x_2} f(x_1)\,dx_1 dx_2 \ldots dx_{n-1} dx_n \tag{6.6.5}$$

From relations (6.6.1) and (6.6.5), we have

$$\int_a^x \int_a^{x_n} \ldots\ldots \int_a^{x_3} \int_a^{x_2} f(x_1)\,dx_1 dx_2 \ldots dx_{n-1} dx_n$$

$$= \frac{1}{(n-1)!} I_n(x) = \frac{1}{(n-1)!}\int_a^x (x-\eta)^{n-1} f(\eta)\,d\eta$$

This may be represented as the result of integrating the function f from a to x and then integrating $(n-1)$ times. We have

https://doi.org/ 10.2478/9788366675575-006

$$\int_a^x f(\eta)\,d\eta^n = \int_a^x \frac{(x-\eta)^{n-1}}{(n-1)!} f(\eta)\,d\eta \cdot \qquad (6.6.6)$$

6.7. Relation between Differential and Integral equation:

There is a fundamental relationship between integral equation and ordinary and partial differential equations with given initial values.

Consider the differential equation of nth order is

$$\frac{d^n y}{dx^n} + a_1(x)\frac{d^{n-1}y}{dx^{n-1}} + a_2(x)\frac{d^{n-2}y}{dx^{n-2}} + ... + a_n(x)y = F(x) \qquad (6.7.1)$$

With continuous coefficients $a_i(x), i = 1,2,...,n$.

The initial conditions are prescribed as follows

$$y(0) = c_0, y'(0) = c_1, y''(0) = c_2, ..., y^{n-1}(0) = c_{n-1} \qquad (6.7.2)$$

Where the prime denotes differentiation with respect to x .

Consider $\frac{d^n y}{dx^n} = \phi(x)$

By integrating and using the integral conditions (6.7.2), we have

$$\frac{d^{n-1}y}{dx^{n-1}} = \int_0^x \phi(\xi)\,d\xi^2 + c_{n-1}$$

$$\frac{d^{n-2}y}{dx^{n-2}} = \int_0^x \phi(\xi)\,d\xi^2 + c_{n-1}x + c_{n-2}$$

https://doi.org/ 10.2478/9788366675575-006

$$y=\int_0^x \phi(\xi)d\xi^n + c_{n-1}\frac{x^{n-1}}{(n-1)!}+c_{n-2}\frac{x^{n-2}}{(n-2)!}+.......+c_0 \quad (6.7.3)$$

Where $\int_0^x \phi(\xi)d\xi^n$ represents for a multiple integral of order n.

From the relation (6.7.3) and (6.7.1), we obtain

$$\phi(x)+a_1(x)\int_0^x \phi(\xi)d\xi + a_2(x)\int_0^x \phi(\xi)d\xi^2$$

$$+........+a_n(x)\int_0^x \phi(\xi)d\xi^n \quad (6.7.4)$$

$$=F(x)+\sum_{i=1}^{n} c_i\psi_i(x)$$

Where

$$\psi_i(x)=a_i(x)+\frac{x}{1!}a_{i+1}(x)+........+\frac{x^{n-1}}{(n-1)!}a_n(x) \quad (6.7.5)$$

$$\phi(x)+\int_0^x \left[a_1(x)+a_2(x)(x-\xi)+........+a_n(x)\frac{(x-\xi)^{n-1}}{(n-1)!}\right]\phi(\xi)d\xi \quad (6.7.6)$$
$$=G_i(x)$$

Where $G_i(x)=F(x)+\sum_{i=1}^{n} c_i\psi_i(x)$

The equation (6.7.6) represents the non-homogenous Volterra integral equation of the second kind. □

https://doi.org/ 10.2478/9788366675575-006

6.8. Particular case:

Consider the linear differential equation of second order

$$\frac{d^2y}{dx^2}+a_1(x)\frac{dy}{dx}+a_2(x)y=F(x) \tag{6.8.1}$$

With initial conditions

$$y(0)=c_0,\ y'(0)=c_1 \tag{6.8.2}$$

Consider $\frac{d^2y}{dx^2}=\phi(x)$

By integrating and using the initial condition (6.8.2), we have

$$\frac{dy}{dx}=\int_0^x \phi(\xi)d\xi+c_1 \tag{6.8.3}$$

And

$$y=\int_0^x (x-\xi)\phi(\xi)d\xi+c_1x+c_0 \tag{6.8.4}$$

The given differential equation reduces to

$$\phi(x)+a_1(x)\left[\int_0^x \phi(\xi)d\xi+c_1\right]+a_2(x)\left[\int_0^x (x-\xi)\phi(\xi)d\xi+c_1x+c_0\right]$$
$$=F(x)$$

Or

$$\phi(x)+\int_0^x\left[a_1(x)+a_2(x)(x-\xi)\right]\phi(\xi)d\xi$$
$$=F(x)-c_1a_1(x)-c_1xa_2(x)-c_0a_2(x)$$

Or $\phi(x)=f(x)+\lambda\int_0^x K(x,\xi)\phi(\xi)d\xi$ (6.8.5)

Where $K(x,\xi)=a_1(x)+a_2(x)(x-\xi)$, $\lambda=-1$

$f(x)=F(x)-c_1a_1(x)-c_1xa_2(x)-c_0a_2(x)$ (6.8.6)

Which represents the Volterra integral equation of the second kind.

Similarly, the BVP in ODE lead to Fredholm Integral equation.

Example-8

Show that the function $\phi(x)=\left(1+x^2\right)^{-\frac{3}{2}}$ is a solution of the Volterra Integral equation

$$\phi(x)=\frac{1}{1+x^2}-\int_0^x\frac{\xi}{1+x^2}\phi(\xi)d\xi$$

Solution:

On substituting the function $\phi(x)=\left(1+x^2\right)^{-\frac{3}{2}}$ in the given equation,

https://doi.org/ 10.2478/9788366675575-006

We have

$$\frac{1}{1+x^2} - \int_0^x \frac{\xi}{1+x^2} \frac{1}{\left(1+\xi^2\right)^{\frac{3}{2}}} d\xi$$

$$= \frac{1}{1+x^2} + \frac{1}{1+x^2} \left[\frac{1}{\left(1+\xi^2\right)^{\frac{3}{2}}} \right]_0^x$$

The substitution $\phi(x)$ reduces the given equation to identity concerning x, Thus

$\phi(x) = \left(1+x^2\right)^{-\frac{3}{2}}$ is a solution of the integral equation. □

6.9. The Fredholm Alternatives:

The Fredholm alternative analyses the solution of the non-homogenous Fredholm equation

$$y(x) = f(x) + \lambda \int_a^b K(x,t) y(t) dt \ , \left(x \in [a,b]\right)$$

Where $f:[a,b] \to R$ and $k:[a,b]^2 \to R$ are continuous and λ is a constant. We shall only establish it in the special and useful case when $K = K(x,t)$ is degenerate (or finite rank), that is when

$$K(x,t)=\sum_{j=1}^{n} g_j(x)h_j(t),\ (x,t\in[a,b])$$

and

$g_j:[a,b]\to R$, $h_j:[a,b]\to R$ are continuous $(j=1,2,....,n)$.

Note that the summation is a finite one. □

A simple case:

To motivate both the statement and proof of the Fredholm Alternatives theorem, we now discuss an equation with perhaps the simplest degenerate kernels $K(x,t)$ of any interest: namely

$$K(x,t)=g(x)h(t),\ (x,t\in[a,b])$$

Where g and h are continuous on $[a,b]$ and neither of which is identically zero on $[a,b]$.

The equation is

$$y(x)=f(x)+\lambda\int_a^b g(x)h(t)y(t)dt=f(x)+\lambda Xg(x)\quad (N)$$

Where $X=\int_a^b h(t)y(t)dt$.

We note immediately that the constant X depends on y.

https://doi.org/ 10.2478/9788366675575-006

We discuss the equation in terms of its (non-homogenous) transpose

$$y(x)=f(x)+\lambda\int_a^b K(t,x)y(t)\,dt$$

$$=f(x)+\lambda\int_a^b g(t)h(x)y(t)\,dt \qquad (N^T)$$

$$=f(x)+\lambda Yh(x)$$

Where $Y=\int_a^b g(t)y(t)\,dt$.

The corresponding homogenous equation

$$y(x)=\lambda\int_a^b g(x)h(t)y(t)\,dt=\lambda X\,g(x) \qquad (H)$$

and that equation's (homogenous) transpose

$$y(x)=\lambda\int_a^b g(t)h(x)y(t)\,dt=\lambda Yh(x) \qquad (H^T)$$

Multiplying (N) by $h(x)$ and integrating with respect to x gives

$$\left(1-\lambda\int_a^b gh\right)X=\int_a^b fh \qquad (6.9.1)$$

Similarly, consideration of $\left(N^T\right)$ results in

https://doi.org/ 10.2478/9788366675575-006

$$\left(1-\lambda\int_a^b gh\right)Y=\int_a^b fg \tag{6.9.2}$$

Provided $\lambda\int_a^b gh\neq 1$, these equations allow the determination of (unique) X and Y, and hence unique continuous solutions.

$$y(x)=f(x)+\frac{\lambda\int_a^b fh}{1-\lambda\int_a^b gh}g(x),\ \ (x\in[a,b])$$

Of (N) and

$$y(x)=f(x)+\frac{\lambda\int_a^b fg}{1-\lambda\int_a^b gh}h(x),\ (x\in[a,b])\ \text{ of }\ \left(N^T\right).$$

However, if $\lambda\int_a^b gh=1$, then neither (N) nor $\left(N^T\right)$ permits a unique solution.

Indeed from (6.9.1), (N) cannot then have a solution unless

$$\int_a^b fh=0 \tag{6.9.3}$$

and from (6.9.2), $\left(N^T\right)$ cannot have a solution unless

$\int_a^b fg = 0$.

Now, every solution of (H), from its very statement, must be of the form

$$y(x) = cg(x) \quad ,\left(x \in [a,b]\right) \tag{6.9.4}$$

Where c is a constant.

Further, if $y = cg$ on $[a,b]$, for some constant c and $\lambda \int_a^b gh = 1$.

Then

$$\begin{aligned} \lambda X g(x) &= c\lambda g(x) \int_a^b h(t) g(t) dt \\ &= cg(x) \\ &= y(x) \end{aligned}$$

For each $x \in [a,b]$ and so (6.9.4) is the (only) solution of (H) .

Similarly, If $\lambda \int_a^b gh = 1$, Then $y(x) = dh(x) . \left(x \in [a,b]\right)$

Where d is a constant is the (only) solution of $\left(H^T\right)$.

https://doi.org/ 10.2478/9788366675575-006

Thus if there does not exist a unique solution to (N) (Where occurs here if $\lambda\int_a^b gh = 1$), then there are non-zero solutions of (H) and (H^T) and there can only be solution of (N) provided the integral of f times the solution of (H^T) is zero.

In the case of this particularly simple kernel and when the "Consistently *condition*" $\int_a^b fh = 0$ is met, it is clear (by simple substitution) that $y = f$ is a particular solution of (N). Further, if $y = Y$ is any solution of (N), then $y = Y - f$ is a solution of (H).

So provided the consistency condition is met and there is no unique solution of (N), the complete solution of (N) is given, with arbitrary constant c,

By $y(x) = f(x) + cg(x) \;, (x \in [a,b])$

(a particular solution plus the complete solution of (H))

Example: -8

https://doi.org/ 10.2478/9788366675575-006

Find the complete solution to each of the following Fredholm equations

(a) $y(x)=x+\lambda\int_0^1 e^{x+t}y(t)\,dt\ ,\ (x\in[0,1])$

(b) $y(x)=x-\frac{2e^x}{e^2-1}+\lambda\int_0^1 e^{x+t}y(t)\,dt,(x\in[0,1])$

Solution:

(a) Consider the first equation

Here the kernel $K(x,t)=e^{x+t}=e^x e^t$ is symmetric, and so

$(N)=(N^T)\ ,\ \ y(x)=x+\lambda Xe^x\ ,$

Where $X=\int_0^1 e^t y(t)\,dt$

Since $\int_0^1 xe^x dx=1\ ,\ \int_0^1 e^{2x}dx=\frac{1}{2}(e^2-1)$

(We always find it useful to carry out such manipulation first),

Multiplying (N) by e^x and integrating gives

$$\left(1-\lambda\frac{(e^2-1)}{2}\right)X=1 \qquad (6.9.5)$$

https://doi.org/ 10.2478/9788366675575-006

So provided $\lambda \neq \frac{2}{(e^2-1)}$ and substituting for X , (N) has the unique continuous solution

$$y(x) = x + \frac{2\lambda}{2-(e^2-1)\lambda} e^x, (x \in [0,1])$$

When $\lambda = \frac{2}{(e^2-1)}$

Equation (6.9.5) shows that there can be no solution to (N).

That is the end of the story for **(a)**.

(b) We have, $(N) = (N^T)$, $y(x) = x - \frac{2e^x}{e^2-1} + \lambda X e^x$,

Where $X = \int_0^1 e^t y(t) dt$

Corresponding to (6.9.5), there is

$$\left(1 - \lambda \frac{(e^2-1)}{2}\right) X = 0 \qquad (6.9.6)$$

and hence provided $\lambda \neq \frac{2}{(e^2-1)}$, (N)

has the unique continuous solution

$$y(x) = x - \frac{2e^x}{e^2-1}, \quad (x \in [0,1])$$

https://doi.org/ 10.2478/9788366675575-006

Where c, c' are arbitrary (though, of course, related) constants.

For both (a) and (b), the homogenous /transposed homogenous equation is $(H)=(H^T)$.

$$y(x)=\lambda Xe^x$$

having the unique solution $y\equiv 0$, when $\lambda\neq\frac{2}{(e^2-1)}$ and when $\lambda=\frac{2}{(e^2-1)}$.

The complete solution $y(x)=ce^x$,

Where c is an arbitrary constant.

So the consistency condition

$\int_a^b fh=0$ in the above discussion is not met in case **(a)**, as

$\int_0^1 xe^x dx\neq 0$, whereas it is met.

In case (b), where

$$\int_0^1\left(x-\frac{2e^x}{e^2-1}\right)e^x dx=0$$

These correspond to the "*right-hand sides*" of (6.9.5) and (6.9.6) being 1 and 0, respectively.

Example-9

Solve the equation

$$y(x)=(e-1)x^n - x+\lambda\int_0^1 xte^{x^3+t^3}y(t)\,dt,\ (x\in[0,1])$$

For the cases $n=1$ and $n=4$.

6.10. The Fredholm alternative theorem

The theorem analyses the solutions of the equation

$$y(x)=f(x)+\lambda\int_a^b K(x,t)y(t)\,dt,\ (x\in[a,b]) \qquad \text{(N)}$$

in terms of the solutions of the corresponding equations

$$y(x)=f(x)+\lambda\int_a^b K(t,x)y(t)\,dt,\ (x\in[a,b]) \qquad \left(N^T\right)$$

$$y(x)=\lambda\int_a^b K(x,t)y(t)\,dt,\ (x\in[a,b]) \qquad (H)$$

$$y(x)=\lambda\int_a^b K(t,x)y(t)\,dt,\ (x\in[a,b]) \qquad \left(H^T\right)$$

https://doi.org/ 10.2478/9788366675575-006

Where $f:[a,b]\rightarrow R$ and the kernel $K:[a,b]^2\rightarrow R$ are continuous and λ is a constant.

Theorem-6.1 (The Fredholm alternative)

For each (fixed) λ precisely, one of the following statements is true:

(F1) The equation (N) possesses a unique continuous solution and, in particular, $f\equiv 0$ on $[a,b]$ implies that $y\equiv 0$ on $[a,b]$. In this case, $\left(N^T\right)$ also possess a unique continuous solution.

(F2) The equation (H) possesses a finite maximal linearly independent set of, say r, continuous solutions $y_1, y_2, \ldots\ldots, y_r$ $(r>0)$.

In this case, $\left(H^T\right)$ also possess a maximal linearly independent set of r-continuous solutions $z_1, z_2, \ldots\ldots, z_r$ and (N) has solutions if and only if the consistency conditions

$$\int_a^b f(x)z_k(x)dx=0,\ (k=1,2,\ldots.r)$$

are all met what they are the complete solution of (N) is given by

$$y(x)=g(x)+\sum_{i=1}^{r}c_iy_i(x),\ (x\in[a,b])$$

Where $c_1,c_2,......,c_r$ are arbitrary constants and $g:[a,b]\to R$ is any continuous solution of (N).

The above is true for the general kernel, but when $K=K(x,t)$ is the degenerate kernel is given by

$$K(x,t)=\sum_{j=1}^{n}g_j(x)h_j(t)\quad (x,t\in[a,b]) \tag{D}$$

There are at most n-values of λ at which (F_2) occurs.

Proof of Theorem-6.1

When K is given by (D), we may assume that each of the sets of functions $(g_1,g_2,.....,g_n),(h_1,h_2,.....,h_n)$ is linearly independent on $[a,b]$. Otherwise, we may express each of their elements is linearly independent subsets to reach the required form.

(For instance, if $(g_1,g_2,.....,g_n)$ is linearly independent but h_n is dependent on the linearly independent subset $(h_1,h_2,.....,h_{n-1})$ and

https://doi.org/ 10.2478/9788366675575-006

$$h_n(t)=\sum_{j=1}^{n} d_j h_j(t),\ (t\in[a,b])$$

Where d_j is a constant $(j=1,2,.....,n)$, then

$$K(x,t)=\sum_{j=1}^{n-1} g_j(x)h_j(t)+g_j(x)\sum_{j=1}^{n-1} d_j h_j(t)$$

$$=\sum_{j=1}^{n-1}\left(g_j(x)+d_j g_j(x)\right)h_j(t)$$

For each x and each t in $[a,b]$.

The reader may easily verify that each of the sets

$(g_1+d_1g_n+.......+d_{n-1}g_n),(h_1,h_2,.....,h_{n-1})$

is linearly independent on $[a,b]$.

For K given by (D), the equation(N), (N^T),(H), (H^T) may be written as

$$y(x)=f(x)+\lambda\sum_{j=1}^{n} X_j g_j(x),$$

Where $X_j=\int_a^b h_j(t)y(t)dt,(j=1,2,...,n)$

(N_1^T) $y(x)=f(x)+\lambda\sum_{j=1}^{n} Y_j h_j(x),$

https://doi.org/ 10.2478/9788366675575-006

Where $Y_j = \int_a^b g_j(t)\,y(t)\,dt$, $(j = 1,2,...,n)$.

$$(H_1)\ y(x) = \lambda \sum_{j=1}^{n} X_j g_j(x)$$

$$(H_1^T)\ y(x) = \lambda \sum_{j=1}^{n} Y_j h_j(x)$$

For x in $[a,b]$. So, a solution of (N) is determined once we know the values of the X_j's (which are, of course, defined in terms of the unknown quantity y).

First, let us note that $\lambda = 0$ corresponds to the unique continuous solution $y = f$ of both (N) and $(N^T) : (F_1)$ occurs.

From this point in the proof, we assume that $\lambda \neq 0$.

We now convert the problem to one of algebra involving a system of simultaneous linear equations.

Multiplying (N_1) through by $h_i(x)$ and integrating concerning x gives on slight re-arrangement,

$$\mu X_i - \sum_{j=1}^{n} a_{ij} X_j = b_i, (i = 1,2,.....,n)$$

Once we define

https://doi.org/ 10.2478/9788366675575-006

$$\mu=\frac{1}{\lambda}, a_{ij}=\int_a^b g_j(x)h_i(x)dx, b_i=\mu\int_a^b f(x)h_i(x)dx$$

These n equations in the X_i's may be rewritten in the compact form

(N_2) $(\mu I-A)X=b$

Where I is the identity matrix, A is the matrix $A=(a_{ij})$, and $X=(X_i)$ $(b=b_i)$ are column vectors.

Similarly, $(N_1^T),(H_1),(H_1^T)$ working in turn with and we have

(N_2^T) $\quad(\mu I-A)^T Y=(\mu I-A^T)Y=c$

Where $A^T=(a_{ji})$ is the transpose of A, $Y=(Y_i),c=(c_i)$ are column vectors and $c_i=\mu\int_a^b f(x)g_i(x)dx, (i=1,2,....,n)$.

(H_2) $(\mu I-A)X=0$

and (H_2^T) $(\mu I-A)^T Y=0$

As we know, If Rank $Rank(\mu I-A)=n,(N_2)$ and (N_2^T) have unique solutions $X=(X_i)$ and $Y=(Y_i)$.

https://doi.org/ 10.2478/9788366675575-006

Substitution of these values in (N_1) and (N_1^T) gives unique solutions to (N) and (N^T) in this case when (F_1) occurs.

We are left with establishing that, when $1 \leq Rank(\mu I - A) = n - r \leq n - 1$, (F_2) must occur.

In this case, both (H_2) and (H_2^T) have maximal linearly independent sets $X^k = X_i^k$ and $Y^k = Y_i^k, k = 1, 2, ..., r$ consisting of r-elements.

Put $y_k(x) = \lambda \sum_{j=1}^{n} X_j^k g_j(x)$

And $z_k(x) = \lambda \sum_{j=1}^{n} Y_j^k h_j(x)$

For each x in $[a,b]$ and $k = 1, 2, ..., r$.

Each y_k is a solution of (H_1) and each z_k a solution of (H_1^T).

We claim now that the set $(y_k : k = 1, 2, ..., r)$ is linearly independent on $[a,b]$.

Suppose that $\sum_{k=1}^{r} c_k y_k = 0$

For some constants $c_1, c_2, ..., c_r$.

https://doi.org/ 10.2478/9788366675575-006

Then $\lambda\sum_{j=1}^{n}\left(\sum_{k=1}^{r}c_kX_j^k\right)g_j=0$

But the g_j's are linearly independent;

So $\sum_{k=1}^{r}c_kX_j^k=0$,for $j=1,2,.....,n$.

That is $\sum_{k=1}^{r}c_kX^k=0$

Since the X^k's are also linearly independent $c_k=0$ for each $k=1,2,...,r$ and the Y^k's are thus independent also.

Similarly, $(z_k:k=1,2,...,r)$ is linearly independent.

These Linearly independent sets $(y_k),(z_k)$ cannot be enlarged whilst remaining independent for the argument above could then be reversed for the augmented set to contradict the maximality of the independent sets (X^k) and (Y^k).

Thus, we have thus shown that maximal linearly independent sets of solutions (H) and (H^T) have the same finite non zero number of elements when (F_1) does not occur.

https://doi.org/ 10.2478/9788366675575-006

It remains to consider (N) in this case (F_2). The proposition of section 8.2 asserts in this context that (N_2) has a solution if and only if

$b^T Y^k = 0$ for $k = 1, 2, \ldots, r$.

That is employing the definitions of b and $(Y^k) = (Y_j^k)$, and noting that Y^k corresponds to the solution z_k of (H_1^T), if and only if

$$\sum_{j=1}^{n}\left(\int_a^b f(x)h_j(x)dx\right)\left(\int_a^b g_j(t)z_k(t)dt\right)=0 \quad \text{for } k=1,2,\ldots,r.$$

This is equivalent to

$$\int_a^b\left(\int_a^b\left(\sum_{j=1}^{n} h_j(x)g_j(t)\right)z_k(t)dt\right)f(x)dx=0$$

for $k = 1, 2, \ldots, r$.

Using again the fact that z_k is a solution of (H^T), the necessary and sufficient condition for (N) to have solution reduces to

$$\int_a^b z_k(x)f(x)dx=0 \quad \text{for} \quad k=1,2,\ldots,r.$$

https://doi.org/ 10.2478/9788366675575-006

Suppose that g is a particular continuous solution of (N) (existing because the consistency conditions are met). Then if y is any other solution of (N), $y-g$ is a solution of (H) and hence expressible as a linear combination of the elements of the maximal independent set of solutions (y_k):

$$y-g=\sum_{i=1}^{r} c_i y_i$$

For some constants $c_1, c_2, \ldots\ldots, c_r$. The Fredholm Alternative theorem is thus established for degenerate kernels.

Note:

The reader will want to distinguish between the eigen values λ corresponding to non-zero solutions of (H) in the case (F_2) and the eigenvalues $\mu=\frac{1}{\lambda}$, that satisfy the matrix equation (H_2).

A worked example:

Example-10

Solve the integral equation

https://doi.org/ 10.2478/9788366675575-006

$$y(x) = f(x) + \lambda \int_0^{2\pi} \sin(x+t)\, y(t)\, dt \ , \ (x \in [0, 2\pi])$$

In the two cases:

(a) $f(x) = 1, \ (x \in [0, 2\pi])$

(b) $f(x) = x, \ (x \in [0, 2\pi])$

The equation may be written $(N) = (N^T)$

$$y(x) = f(x) + \lambda X_1 \sin x + \lambda X_2 \cos x$$

Where $X_1 = \int_0^{2\pi} y(t) \cos t dt \, , X_2 = \int_0^{2\pi} y(t) \sin t dt$

Note that

$$\int_0^{2\pi} \cos^2 x dx = \int_0^{2\pi} \sin^2 x dx = \pi$$

$$\int_0^{2\pi} x \sin x dx = -2\pi$$

$$\int_0^{2\pi} x \cos x dx = \int_0^{2\pi} \cos x dx$$

$$= \int_0^{2\pi} \sin x dx$$

$$= \int_0^{2\pi} \sin x \cos x dx = 0$$

https://doi.org/ 10.2478/9788366675575-006

Multiplying (N) through by $\cos x$ and integrating concerning x and carrying out the same operation with $\sin x$ giving

$$\begin{aligned} X_1 - \lambda\pi X_2 &= \int_0^{2\pi} f(x)\cos x dx \\ -\lambda\pi X_1 + X_2 &= \int_0^{2\pi} f(x)\sin x dx \end{aligned} \qquad (6.10.1)$$

When the determinant of the coefficients of X_i's is nonzero,

$$\begin{vmatrix} 1 & -\lambda\pi \\ -\lambda\pi & 1 \end{vmatrix} = 1 - \lambda^2\pi^2 \neq 0$$

These equations have the unique solutions

$$X_1 = \frac{1}{1-\lambda^2\pi^2}\int_0^{2\pi} f(x)(\cos x + \lambda\pi \sin x)dx$$

$$X_2 = \frac{1}{1-\lambda^2\pi^2}\int_0^{2\pi} f(x)(\sin x + \lambda\pi \cos x)dx$$

In case **(a)**, we thus have the solution

$$y(x) = 1, \ \ x \in [0, 2\pi]$$

and in case the solution

$$y(x) = x - \frac{2\pi\lambda}{1-\lambda^2\pi^2}(\cos x + \lambda\pi \sin x),$$

Provided $1-\lambda^2\pi^2 \neq 0$.we have dealt with (F_1) .

The case (F_2)can only occur when $\lambda = \pm\pi^{-1}$.It is easy to check that the corresponding homogenous equation $(H)=(H^T)$ $y(x)=\lambda X_1 \sin x + \lambda X_2 \cos x$ only has the solutions.

$y(x)=c(\sin x+\cos x)$, When $\lambda=\pi^{-1}$, $(x \in [0,2\pi])$

$y(x)=d(\sin x-\cos x)$, When $\lambda=\pi^{-1}$, $(x \in [0,2\pi])$

Where c and d are constants. (To check this, consider the equations (6.10.1) with the right-hand sides zero, as is the case with the homogenous equation)

The consistency conditions which need to meet to allow solutions (N) to existing are therefore

$$\int_0^{2\pi} (\sin x+\cos x) f(x)\,dx = 0 \text{ , when } \lambda=\pi^{-1}.$$

and $\int_0^{2\pi} (\sin x-\cos x) f(x)\,dx = 0$, when $\lambda=-\pi^{-1}$.

In case **(a)** both conditions are fulfilled.

https://doi.org/ 10.2478/9788366675575-006

As $y(x)=f(x)=1$, $(x\in[0,2\pi])$ is a particular solution when either $\lambda=\pi^{-1}$ or $\lambda=-\pi^{-1}$ the complete solution of (N) is when $\lambda=\pi^{-1}$.

$y(x)=1+c(\sin x+\cos x)$, $(x\in[0,2\pi])$

And when $\lambda=-\pi^{-1}$,

$y(x)=1+d(\sin x-\cos x)$, $(x\in[0,2\pi])$

Where c and d are arbitrary constants.

In case **(b)**, neither the condition is met and (N) has, therefore, no solution when either $\lambda=\pi^{-1}$ or $\lambda=-\pi^{-1}$.

To illustrate these points, we consider several examples.

Example-11

Consider the equation

$$y(x)=e^x+\lambda\int_0^1(5x^2-3)t^2y(t)\,dt.$$

Solution:

Here $y(x)=e^x+\lambda c(5x^2-3)$

Where

https://doi.org/ 10.2478/9788366675575-006

$$c = \int_0^1 t^2 y(t)\,dt$$

$$= \int_0^1 t^2 \left(e^t + \lambda c\left(5t^2 - 3\right)\right) dt$$

$$= e - 2$$

For any λ, the given equation has the unique solution

$$y(x) = e^x + \lambda(e-2)\left(5x^2 - 3\right)$$

It may be verified that the homogenous equation

$$y(x) = \lambda \int_0^1 \left(5x^2 - 3\right) t^2 y(t)\,dt$$

has the only trivial solution. □

Example-12

Consider the equation

$$y(x) = 2x + \lambda \int_0^1 \sin \ln x \, y(t)\,dt$$

Solution:

Here, $y(x) = 2x + \lambda c \sin \ln x$

Where $c = \int_0^1 y(t)\,dt = \int_0^1 (2t + \lambda c \sin \ln t)\,dt$

$$\therefore c\left(1 + \frac{\lambda}{2}\right) = 1$$

https://doi.org/ 10.2478/9788366675575-006

If $\lambda \neq -2$, then the equation has a unique solution i.e.

$$y(x) = -2x + \frac{2\lambda}{2+\lambda}\sin \text{lnx}$$

and the corresponding homogenous equation

$$y(x) = -2\int_0^1 \sin \ln x\, y(t)\, dt$$

has the only trivial solution $y(x) = 0$.

If $\lambda = -2$, the given equation has no solution. □

Example-13

$$y(x) = \cos 3x + \lambda \int_0^{\pi} \cos(x+t)\, y(t)\, dt$$

Solution:

Here $y(x) = \cos 3x + \lambda c_1 \cos x - \lambda c_2 \sin x$

Where

$$c_1 = \int_0^{\pi} \cos \text{t}\, y(t)\, dt$$

$$= \int_0^{\pi} \cos \text{t}\left[\cos 3t + \lambda c_1 \, \text{cost} - \lambda c_2 \, \text{sint}\right] dt$$

https://doi.org/ 10.2478/9788366675575-006

$$c_2 = \int_0^{\pi} \sin t\, y(t)\, dt$$

$$= \int_0^{\pi} \sin t\left[\cos 3t + \lambda c_1 \cos t - \lambda c_2 \sin t\right] dt$$

Evaluating these integrals, we find

$$\left(1 - \frac{\lambda\pi}{2}\right)c_1 = 0, \left(1 + \frac{\lambda\pi}{2}\right)c_2 = 0$$

$\therefore$ If $\lambda \neq \pm\frac{2}{\pi}$, $c_1 = c_2 = 0$, then the given equation has a unique solution $y(x) = \cos 3x$.

On the other hand, If $\lambda = \frac{2}{\pi}$, c_1 is arbitrary and $c_2 = 0$. Therefore the given equation has infinite number of solutions $y(x) = \cos 3x + c\cos x$.

If $\lambda = -\frac{2}{\pi}$ the given equation has an infinite number of solutions $y(x) = \cos 3x + c\sin x$. □

Example-14

Obtain the most general solution for

$$y(x) = \lambda \int_0^{2\pi} \sin(x+t)\, y(t)\, dt + f(x)$$

https://doi.org/ 10.2478/9788366675575-006

Where $f(x)=1$ and $f(x)=x$, under the assumption that $\lambda \neq \pm\frac{1}{\pi}$.

Solution:

Here $y(x)=f(x)+\lambda c_1 \sin x+\lambda c_2 \cos x$.

Where $c_1=\lambda\int\limits_0^{2\pi}\cos t\, y(t)\,dt$, $c_2=\lambda\int\limits_0^{2\pi}\sin t\, y(t)\,dt$

$$\therefore c_1-\lambda\pi c_2=\int\limits_0^{2\pi}\mathrm{f}(t)\cos t\,dt$$

$$-\lambda\pi c_1+c_2=\int\limits_0^{2\pi}\mathrm{f}(t)\sin t\,dt$$

If $f(x)=x$,then

$$c_1-\lambda\pi c_2=0$$
$$-\lambda\pi c_1+c_2=-2\pi$$

Solving $c_1=\dfrac{2\lambda\pi^2}{\lambda^2\pi^2-1}$ and $c_2=\dfrac{2\pi}{\lambda^2\pi^2-1}$

We find $y(x)=x+\dfrac{2\pi\lambda}{\lambda^2\pi^2-1}(\pi\lambda\sin x+\cos x)$.

If $f(x)=1$,we have

$$c_1-\lambda\pi c_2=0$$
$$-\lambda\pi c_1+c_2=0$$

Since $\lambda \neq \frac{1}{\pi}$,we have $c_1 = c_2 = 0$ and the desired solution is

$y(x) = 1$.□

Example-15

Solve $y(x) = \sin x + \lambda \int_0^{2\pi} \cos(x+t) y(t) dt$

Solution:

Here $y(x) = \sin x + \lambda c_1 \cos x - \lambda c_2 \sin x$

Where

$$c_1 = \int_0^{2\pi} \left[(1 - \lambda c_2) \sin t \cos t + \lambda c_1 \cos^2 t \right] dt$$

$$c_2 = \int_0^{2\pi} \left[(1 - \lambda c_2) \sin^2 t + \lambda c_1 \cos t \sin t \right] dt$$

Evaluating these integrals, it finds

$$(1 - \lambda\pi) c_1 = 0$$
$$(1 + \lambda\pi) c_2 = \pi$$

If $\lambda \neq \pm\frac{1}{\pi}$,then $c_1 = 0$,$c_2 = \frac{\pi}{1+\lambda\pi}$.

Then $y(x) = \frac{\sin x}{1+\lambda\pi}$.

https://doi.org/ 10.2478/9788366675575-006

If $\lambda = \frac{1}{\pi}$,then c_1 is arbitrary and $c_2 = \frac{\pi}{2}$.

$y(x) = \frac{1}{2}\sin x + c\cos x$, where c is arbitrary.

If $\lambda = -\frac{1}{\pi}$, then there is no solution. □

https://doi.org/ 10.2478/9788366675575-006

Exercises

Investigate for solvability the following integral equations for different value of the parameter λ.

1. $y(x) = x + \lambda \int_{-1}^{1} xe^{t}\, y(t)\, dt$

Ans. $y(x) = \dfrac{ex}{e-2\lambda}, \lambda \neq \dfrac{e}{2}$

No solution for $\lambda = \dfrac{e}{2}$.

2. $y(x) = x + \lambda \int_{0}^{2\pi} |x-\pi|\, y(t)\, dt$

Ans. $y(x) = x + \dfrac{2\pi^2 \lambda |x-\pi|}{1-\pi^2\lambda}, \lambda \neq \dfrac{1}{\pi^2}$

No solution for $\lambda = \dfrac{1}{\pi^2}$.

3. $y(x) = 1 - 2x + \lambda \int_{0}^{1} \left(2xt - 4x^2\right) y(t)\, dt$

Ans. $y(x) = \dfrac{3x\left(2\lambda^2 x - 2\lambda^2 - 5\lambda - 6\right) + (\lambda+3)^2}{(\lambda+3)^2}, \lambda \neq -3$

No solution for $\lambda = -3$.

https://doi.org/ 10.2478/9788366675575-006

4. $y(x)=x^3-x+\lambda\int_{-1}^{1}\left(x^2-2xt\right)y(t)\,dt$

Ans.
$$y(x)=x^2-\frac{3x(4\lambda+5)}{5(4\lambda+3)}\ ,\lambda\neq\frac{3}{2},\lambda\neq-\frac{3}{4}$$
$$=x^3-\frac{11x}{15}+cx^2\ ,\lambda=\frac{3}{2}$$

No solution for $\lambda=-\dfrac{3}{4}$.

5. $y(x)=\sin x+\lambda\int_{0}^{2\pi}\left(\frac{1}{\pi}\cos x\cos t+\frac{1}{\pi}\sin 2x\sin 2t\right)y(t)\,dt$

Ans.
$$y(x)=\sin x\ ,\ \lambda\neq 1$$
$$=\sin x+c_1\cos x+c_2\sin 2x\ \ \lambda=1$$

6. Prove that the equation

$y(x)=\dfrac{1}{\pi}\int_{0}^{2\pi}\sin(x+t)y(t)\,dt+f(x)$ possess no solution when $f(x)=x$,but that it possesses infinitely many solutions when $f(x)=1$. Determine all such solutions.

Ans. $y=1+c(\cos x+\sin x)$.

https://doi.org/ 10.2478/9788366675575-007

Chapter-7

The method of undetermined coefficients

7.1. Approximation methods of undetermined coefficients

Other numerical methods for obtaining approximate solutions of integral equations also generally include reducing the problem to considering a finite set of algebraic equations.

In particular, the solution of the equation

$$y(x)=f(x)+\lambda\int_a^b K(x,t)y(t)dt \tag{7.1.1}$$

may be approximated by a linear combination of n-suitably chosen functions $\phi_1,\phi_2,......,\phi_n$ of the form

$$y(x)\approx\sum_{k=1}^{n}c_k\phi_k(x) \tag{7.1.2}$$

Where the n constants of combination are to be determined in such a way that equation (7.1.1) is satisfied as nearly possible (in some sense) by (7.1.2) over the interval (a,b).

The requirement that (7.1.2) approximately satisfy (7.1.1) takes the form

$$\sum_{k=1}^{n}c_k\phi_k(x)\approx f(x)+\lambda\sum_{k=1}^{n}c_k\int_a^b K(x,t)\phi_k(t)dt \tag{7.1.3}$$

https://doi.org/ 10.2478/9788366675575-007

With the convenient abbreviation

$$\Phi_k(x)=\int_a^b K(x,t)\phi_k(t)\,dt \tag{7.1.4}$$

This condition becomes merely

$$\sum_{k=1}^{n} c_k\left[\phi_k(x)-\lambda\Phi_k(x)\right]\approx f(x)\quad ,(a\le x\le b) \tag{7.1.5}$$

The coefficients $c_1,c_2,.....,c_n$ are then to be determined by a set of n conditions that tend to reduce the two members of (7.1.5) to equality over the interval (a,b).

Several procedures which lead to such sets of conditions are outlined and illustrated in the sections which follow.

In practical cases, advanced information concerning the nature of the behaviour of the unknown function $y(x)$ is frequently at hand, and the choice of the approximating functions is motivated by this knowledge.

It is frequently convenient to take the approximation in the form of a polynomial of the degree n to identify with x^k . However, if one or more of the end values $y(a)$ and $y(b)$ is known in advance (or obtainable by inspection from the integral equation, as indeed was the case in the example of the

https://doi.org/ 10.2478/9788366675575-007

preceding section), then it may be desirable to take the assumed approximation in the form

$$y(x) \approx \phi_0(x) + \sum_{k=1}^{n} c_k \phi_k(x) \tag{7.1.6}$$

Where ϕ_0 is chosen so that it assumes the known end values, and the remaining ϕ's are made to vanish at the corresponding end or ends of the interval.

The approximation obtained dependability can be judged by comparing the resultant left-hand member of (7.1.5) with the right-hand member.it should be pointed out, however, that situation, unfortunately, exist in which a significant change in the function $y(x)$ may correspond to a slight change in the function

$$y(x) - \lambda \int_a^b K(x,t)\, y(t)\, dt$$

In such cases, the integral equation can be very satisfied over the interval (a,b) by an "approximation", in the sense that the difference between the two members of (7.1.5) then is everywhere small relative to either of those members, but $\tilde{y}(x)$ may differ appreciably from the exact solution $y(x)$ over that interval.

https://doi.org/ 10.2478/9788366675575-007

A somewhat fairer estimate of dependability is obtained by comparing the result of an n-term approximation with the development of an $(n+1)$–term approximation. The same technique determines the constants of combination are determined by the same approach in both cases. □

7.2. The method of collocation

If we introduce the abbreviation

$$\begin{aligned} s_k(x) &= \phi_k(x) - \lambda \Phi_k(x) \\ &\equiv \phi_k(x) - \lambda \int_a^b K(x,t)\phi_k(t)\,dt \end{aligned} \tag{7.2.1}$$

The requirement that (7.1.2) approximately satisfy (7.1.1) can be expressed in the form

$$\sum_{k=1}^{n} c_k s_k(x) \approx f(x) \quad , (a \le x \le b) \tag{7.2.2}$$

A set of n-conditions for determining the n constants of combination can be obtained most simply by requiring that (7.2.2) be equally at n distinct points in the interval (a,b). If we denote these points by x_i $(i = 1,2,.....n)$, then the resultant conditions are then of the form

$$\sum_{k=1}^{n} c_k s_k(x_i) = f(x_i) \quad (i = 1, 2, ...n) \tag{7.2.3}$$

The matrix of the coefficients of $c's$ in this set of equations is then given by

$$S = [s_{ij}] \text{ Where } s_{ij} = s_j(x_i) \tag{7.2.4}$$

That is, the set of equations can be expressed in the matrix form

$$Sc = f \tag{7.2.5}$$

Where $c = \{c_i\}$ and $f = \{f_i\}$.

To illustrate the method, we again consider the integral equation

$$y(x) = x + \lambda \int_0^1 K(x,t) y(t) dt \tag{7.2.6}$$

where $K(x,t)$ is defined by

$$K(x,t) = \begin{cases} x(1-t), & x < t \\ t(1-x), & x > t \end{cases}.$$

For simplicity, we assume a three-term approximation of the polynomial form

$$y(x) \approx c_1 + c_2 x + c_3 x^2 \tag{7.2.7}$$

https://doi.org/ 10.2478/9788366675575-007

With $\phi_1 = 1, \phi_2 = x$ and $\phi_3 = x^2$ with the notation of (7.1.4), there follows by direct integration

$$\Phi_1 = \frac{1}{2}x(1-x),$$
$$\Phi_2 = \frac{1}{6}x(1-x^2), \qquad (7.2.8)$$
$$\Phi_3 = \frac{1}{12}x(1-x^3)$$

Thus, with the notation of (7.2.1), equation (7.2.2) here takes the form

$$c_1\left[1-\frac{1}{2}x(1-x)\right]+c_2\left[x-\frac{1}{6}x(1-x^2)\right] + c_3\left[x^2-\frac{1}{12}x(1-x^3)\right] \approx x \quad ,(0 \le x \le 1)(7.2.9)$$

If we require that this relation be equality at the three points

$x = 0, x = \frac{1}{2}$ and $x = 1$,

we obtain the conditions

$$c_1 = 0$$
$$\frac{7}{8}c_1 + \frac{7}{16}c_2 + \frac{41}{192}c_3 = \frac{1}{2} \qquad (7.2.10)$$
$$c_1 + c_2 + c_3 = 1$$

The solution of this set of equations, with the results rounded to four decimal places, is then given by

https://doi.org/ 10.2478/9788366675575-007

$$c_1 = 0\ , c_2 = 1.2791\ , c_3 = -0.2791 \qquad (7.2.11)$$

So that the desired approximate solution is of the form

$$y(x) \approx 1.2791x - 0.2791x^2 \qquad (7.2.12)$$

7.3. The method of weighting functions:

A second method of obtaining n conditions for the determination of the constants, which is often associated with the name of *Galerkin*, consists of requiring that the difference between the two members of (8) be orthogonal to n linearly independent functions $\psi_i(x)$ $(i = 1,2,.....n)$ over the interval (a,b).

Thus, the conditions obtained in this way are of the form

$$\sum_{k=1}^{n} c_k \int_a^b \psi_i s_k dx = \int_a^b \psi_i f\, dx \quad (i = 1,2,.....n) \qquad (7.3.1)$$

Or equivalently

$Mc = b$, Where

$$m_{ij} = \int_a^b \psi_i s_j dx \text{ and } b_i = \int_a^b \psi_i f\, dx. \qquad (7.3.2)$$

The procedure can be interpreted as weighting the two members of (7.2.2) by each of the functions $\psi_i(x)$ and requiring that the integrals of the weighted members be equal. A particularly convenient choice of the n weighting functions is the set $1, x, x^2, ..., x^{n-1}$.

In this case the graphical representations of the two members of (7.2.2) are required to determine areas with the x-axis which are equal and whose first $n-1$ moments are equal. It is desirable to choose the functions ψ_i as n-members of a complete set of functions, since then the relation (7.2.2) must tend to equality over (a,b) as n increases without limit. It is often convenient to identify the weighting functions ψ_i with the approximating functions ϕ_i.

In the illustration, the application of this procedure to the example considered previously, with the weighting functions $1, x$ and x^2 leads to the conditions

$$\begin{aligned} \frac{11}{12}c_1 + \frac{11}{24}c_2 + \frac{37}{120}c_3 &= \frac{1}{2} \\ \frac{11}{24}c_1 + \frac{14}{45}c_2 + \frac{17}{72}c_3 &= \frac{1}{3} \\ \frac{37}{120}c_1 + \frac{17}{72}c_2 + \frac{321}{1680}c_3 &= \frac{1}{4} \end{aligned} \qquad (7.3.3)$$

https://doi.org/ 10.2478/9788366675575-007

These equations are obtained by multiplying the two members of (7.2.9) successively $1, x$ and x^2 and equating the integrals of the results over (0,1). The solution is found to be

$$c_1 = -0.0088\ , c_2 = 1.2968, c_3 = -0.2798 \qquad (7.3.4)$$

Leading to the approximation

$$y(x) \approx -0.0088 + 1.2968x - 0.2798x^2 . \square$$

7.4. The method of least squares

The accuracy obtained by the procedures of the two preceding sections will be in general, depend upon the choice of appropriate points of collocation or weighting functions. A method that avoids this dependence upon the judgement of the computer is next presented.

In place of requiring that the integral equation be satisfied exactly at several points equal to the number of undermined coefficients, we may need that the integral of the square of the difference between the two members, over (a,b) , be as small as possible. Thus, the basic condition is of the form

$$\int_a^b \left[\sum_{k=1}^{n} c_k s_k(x) - f(x) \right]^2 dx = \text{minimum} \qquad (7.4.1)$$

https://doi.org/ 10.2478/9788366675575-007

With the notation of (7.2.1) and (7.2.2). So that (7.4.1) be satisfied, the derivative of the left-hand member concerning each parameter c_i must vanish so that we must have

$$\int_a^b s_i(x)\left[\sum_{k=1}^{n} c_k s_k(x) - f(x)\right]dx = 0 \quad (i = 1, 2, \ldots\ldots, n) \qquad (7.4.2)$$

These conditions take the form

$$\sum_{k=1}^{n} c_k \int_a^b s_i s_k\, dx = \int_a^b s_i f\, dx \qquad (7.4.3)$$

and hence are equivalent to the requirements (8.3.1) where the weighting functions $\psi_i(x)$ are identified with the functions $s_i(x)$.

Thus, it follows that if the integral equation is to be satisfied as well as possible over (a,b) in the least-squares sense, the weighting functions of the preceding section must be identified with the functions $s_i(x)$.

In many practical cases, the functions are such that the integrations involved in the above equation are not feasible. Therefore, a modification that incorporates most of this method's advantages over the collocation procedure, with only a slight increase in the amount of calculation involved, is now formulated.

If weighted sums of the relevant ordinates approximate the integrals in (7.4.1) and (7.4.2) at N conveniently chosen points, the resultant minimal conditions (7.4.2) take the form

$$\sum_{r=1}^{N} D_r s_i(x_r)\left[\sum_{k=1}^{n} c_k s_k(x_r) - f(x_r)\right] = 0, (i = 1,2,......,n) \qquad (7.4.4)$$

Where the numbers $D_r\ (r = 1,2,......,N)$ are appropriate weighting coefficients associated with the points $x_1, x_2, ..., x_N$ involved in the approximate integration. These conditions can also be expressed in the form

$$\sum_{k=1}^{n} c_k\left[\sum_{r=1}^{N} D_r s_i(x_r) s_k(x_r)\right] = \sum_{r=1}^{N} D_r s_i(x_r) f(x_r), (i = 1,2,......,n) \qquad (7.4.5)$$

Despite the rather formidable appearance of this set of conditions, the coefficients in the set of linear algebraic equations, which it represents, can be obtained very simply by matrix multiplication, as is next shown.

Equation (8.4.5) can be written in the abbreviated form

https://doi.org/ 10.2478/9788366675575-007

$$\sum_{k=1}^{n} c_k p_{ik} = q_i \ ,(i = 1,2,......,n) \tag{7.4.6}$$

Where

$$p_{ik} = \sum_{r=1}^{N} D_r s_{ri} s_{rk} \tag{7.4.7}$$

And

$$q_i = \sum_{r=1}^{N} D_r s_{ri} f_r \tag{7.4.8}$$

and where we have written

$$s_{ij} = s_j(x_i) \tag{7.4.9}$$

Following the notation of (8.2.4). With a change of indices, equation (8.4.7) can be written in the form

$$p_{ij} = \sum_{k=1}^{N} s_{ki} D_k s_{kj} \quad (i, j = 1,2,......,n) \tag{7.4.10}$$

And hence can be expressed in the matrix form

$$P = S^T DS \tag{7.4.11}$$

Where $P = [p_{ij}], S = [s_{ij}] \equiv [s_j(x_i)]$ and $D = [D_i \delta_{ij}]$.

In the same way, we obtain also

$$q = S^T Df \tag{7.4.12}$$

Where $q = \{q_i\}$ and $f = \{f_i\}$.

Further, since D is a diagonal matrix, there follows also

https://doi.org/ 10.2478/9788366675575-007

$$S^T D = (DS)^T$$

So that (7.4.11) and (7.4.12) become

$$P = (DS)^T S \ , \ q = (DS)^T f \qquad (7.4.13)$$

If we notice that S is the matrix of the coefficients of the $c's$ in the equations

$$\sum_{k=1}^{n} c_k s_k (x_i) = f(x_i) \ \ (i = 1, 2, ..., N) \qquad (7.4.14)$$

We see that the matrix of coefficients P in (7.4.5) can be obtained by pre-multiplying the matrix of coefficients in (7.4.13) by the transpose of the matrix DS, and the column of right-hand side members q in (7.4.5) can be obtained, similarly, by pre-multiplying the corresponding column f in (7.4.14) by the same matrix.

This result leads to the following procedure for determining the n linear equations represented by (7.4.5):

1. Choose N points $x_1, x_2, ..., x_N$ in the interval (a,b) and write down the N equations (7.4.14) requiring that the integral equation be satisfied at those points when $y(x)$ is replaced by the approximation (7.1.2).

https://doi.org/ 10.2478/9788366675575-007

2. Denote the $N \times n$ matrix of coefficients in this set of equations by S and form an associated "weighting matrix"

$S^* = DS$

By multiplying the ith row of S by the weighting coefficient D_i associated with the point x_i in an approximate integration scheme involving the N points.

3. Pre-multiply the *augmented matrix* of (7.4.14) by the *transpose* of the weighting matrix S^*. The resultant matrix is the augmented matrix of the required set of n-linear equations, determining the constants $c_1, c_2, \ldots\ldots, c_n$.

We may notice that, since (7.4.5) is homogenous in the $D's$, these weighting coefficients may be multiplied by any convenient common factor in the formation of S^*. Thus if the trapezoidal formula is used, then the successive coefficients are conveniently taken to be merely $\frac{1}{2}, 1, 1, \ldots, 1, 1, \frac{1}{2}$. The elements of the first and last rows of S are divided by two, and the remaining entries are unchanged. Similarly, If Simpson's rule is used, the coefficients can be taken as $\frac{1}{2}, 2, 1, 2, 1, \ldots, 1, 2, \frac{1}{2}$.

https://doi.org/ 10.2478/9788366675575-007

As may be expected, this method leads to a set of equations equivalent to the original set when $N=n$ that is, when the number of chosen points x_i is equal to the number of $c's$ to be determined. However, when $N>n$, it permits us to choose several points greater than n and require that the integral equation be satisfied as nearly as possible at those points, rather than to require that it be satisfied *exactly* at n points. The weighting coefficients D_i weight the squared errors committed to failing to fulfil the equation at the respective points in proportion to the influence of the ordinate at x_i in the integration of the squared error over (a,b).whereas the N equations (7.4.14) are in general incompatible, this procedure affords the "best possible" solution in a least-square sense.

It should be noticed that, by substitution of the calculated $c's$ into the left-hand members of the equations (7.4.14), the difference between the two members of the integral equation can be readily calculated at the N chosen points to indicate the dependability of the solution. In physically motivated problems, it is often clear from the nature of the relevant natural phenomenon that small errors in the integral equation's satisfaction necessarily imply small errors in the unknown functions. However, as was mentioned earlier, this situation does not always exist.

The present procedure differs from the collocation procedure: First, more than n equations are formed initially; second, a weighting matrix must be determined; and third, additional matrix multiplication is involved. Since additional equations would be needed in any case to investigate the degree of satisfaction of the integral equation, this feature involves no other calculation. As was shown, the formation of the weighting matrix need involves only multiplication or division of some aspects in the original coefficient matrix by a factor of two if the formulas of the trapezoidal rule or Simpson's rule are used. The principal source of increased labour is involved in matrix multiplication. However, the relevant operations are particularly well adapted (for example) to the use of automatic desk calculators, each element of the product matrix being determined by a single continuous sequence of machine operations.

When large (so that the matrix S possesses many rows), it is often inconvenient to write the weighting matrix S^* in transposed form. In such cases, it may be preferable to merely write the matrix S^* to the left of the original augmented matrix, without transposing its rows and columns, and to determine the product by column-into-column (rather than row-into-column) multiplication. The element in the $i\,th$ row and $j\,th$ column of the product matrix is then formed from the $i\,th$ column of the first factor and the $j\,th$ column of the second factor.

It is helpful to notice that it follows from (7.4.10) that

$$p_{ji} = p_{ij} \tag{7.4.15}$$

So that the coefficient matrix of the final set of equations is *symmetric*. This means that all elements below the principal diagonal of P need not be calculated directly but may be written down by symmetry once the remaining elements have been determined. The symmetry of the coefficient matrix also permits an appreciable reduction of labour in the actual solution of the corresponding set of equations.

To illustrate the procedure just considered, we again deal with the example of the preceding section.

If we choose the five points $x = 0, \frac{1}{4}, \frac{1}{2}, \frac{3}{4}$ 1 as the points x_i, the five equations corresponding to (8.4.14) are obtained by

https://doi.org/ 10.2478/9788366675575-007

equating the two members of (7.2.9) for those five values of x_i in the form

$$\begin{aligned}
&c_1 = 0 \\
&\frac{29}{32}c_1 + \frac{27}{128}c_2 + \frac{43}{1024}c_3 = \frac{1}{4} \\
&\frac{7}{8}c_1 + \frac{7}{16}c_2 + \frac{41}{192}c_3 = \frac{1}{2} \\
&\frac{29}{32}c_1 + \frac{89}{128}c_2 + \frac{539}{1024}c_3 = \frac{3}{4} \\
&c_1 + c_2 + c_3 = 1
\end{aligned} \tag{7.4.16}$$

If five decimal places are retained in the calculations, the augmented matrix of the required set of three equations is then obtained as follows:

$$\begin{bmatrix} 0.50000 & 0 & 0 \\ 1.81250 & 0.42188 & 0.08398 \\ 0.87500 & 0.43750 & 0.21354 \\ 1.81250 & 1.39062 & 1.05274 \\ 0.50000 & 0.50000 & 0.50000 \end{bmatrix}^T \begin{bmatrix} 1 & 0 & 0 & 0 \\ 0.90625 & 0.21094 & 0.04199 & 0.25000 \\ 0.87500 & 0.43750 & 0.21354 & 0.50000 \\ 0.90625 & 0.69531 & 0.52637 & 0.75000 \\ 1 & 1 & 1 & 1 \end{bmatrix}$$

$$= \begin{bmatrix} 5.05078 & 2.52539 & 1.71700 & 2.75000 \\ 2.52539 & 1.74731 & 1.34312 & 1.86718 \\ 1.71700 & 1.34312 & 1.10326 & 1.41732 \end{bmatrix} \tag{7.4.17}$$

The second factor in the product is merely the augmented matrix of (7.4.16). In forming the weighting matrix which precedes it, we have used the weighting

https://doi.org/ 10.2478/9788366675575-007

coefficients $\frac{1}{2}, 2, 1, 2, \frac{1}{2}$ corresponding to the formula of Simpson's rule.

The corresponding set of equations

$$\begin{aligned} 5.05078c_1 + 2.52539c_2 + 1.71700c_3 &= 2.75000 \\ 2.52539c_1 + 1.74731c_2 + 1.34312c_3 &= 1.86718 \\ 1.71700c_1 + 1.34312c_2 + 1.10326c_3 &= 1.41732 \end{aligned} \tag{7.4.18}$$

Then is found to possess the solution

$$c_1 = -0.0079\ ,\ c_2 = 1.2939,\ c_3 = -0.2783 \tag{7.4.19}$$

Leading to the approximation

$$y(x) \approx -0.0079 + 1.2939x - 0.2783x^2 \tag{7.4.20}$$

7.5. Approximation of the kernel

It is sometimes convenient to approximate the *kernel* of a Fredholm integral equation by a polynomial in x and t or by a *separable kernel* of a more general form and to solve the resulting approximate equation by the methods of that section. Thus, for example, the kernel could be approximated by a three-parameter polynomial, of the form $A_1 + A_2x + A_3x^2$ or of the more appropriate form $x(1-x)\left(B_1 + B_2x + B_3x^2\right)$, where the $A's$ or $B's$ are determined as functions of t by three-point collocation, the use of proper weighting functions, or the use of least square techniques.

To illustrate the procedure, we assume a crude approximation of the form

$$K(x,t) \approx Bx(1-x) \tag{7.5.1}$$

Noticing that the approximation is exact at the end points $x=0$ and $x=1$ .we determine the coefficient B in such a way that the integral of the kernel over $(0,1)$ is equal to the integral of its approximation over that interval:

$$\int_0^1 K(x,t)\,dx = B\int_0^1 x(1-x)\,dx \tag{7.5.2}$$

The direct calculation then gives the determination

$$B = 3t(1-t)$$

And the introduction of the corresponding approximate kernel leads to the approximating integral equation

$$y(x) = x + 3x(1-x)\int_0^1 t(1-t)\,y(t)\,dt \tag{7.5.3}$$

Following the method of previous section, we introduce the abbreviation

$$c = \int_0^1 t(1-t)\,y(t)\,dt \tag{7.5.4}$$

https://doi.org/ 10.2478/9788366675575-007

And rewrite (7.5.3) in the form

$$y(x) = x + 3cx(1-x) \tag{7.5.5}$$

To determine c, we multiply the equal members of (7.5.5) by $x(1-x)$ and integrate the results over $(0,1)$, to obtain the condition

$$c = \int_0^1 t^2(1-t)\,dt + 3c\int_0^1 t^2(1-t)^2\,dt$$

and the evaluation

$$c = \frac{5}{54} \tag{7.5.6}$$

Hence the desired approximate solution (7.5.5) is obtained in the form

$$\begin{aligned} y(x) &\approx x + \frac{5}{18}x(1-x) \\ &= 1.2778x - 0.2778x^2 \end{aligned}$$

More generally, it is easily seen that a kernel approximation of the particular form

$$K(x,t) \approx xt(1-x)(1-t)\left(a_1 + a_2xt + a_3x^2t^2 + \dots\dots\right) \tag{7.5.7}$$

Would lead to an approximate solution of the form

$$y(x) \approx x + x(1-x)\left(c_1 + c_2x + c_3x^2 + \dots\dots\right) \tag{7.5.8}$$

□

https://doi.org/ 10.2478/9788366675575-007

Exercises

1. Solve the linear Fredholm integral equation,

$$y(x)=x+\int_0^1 (1+x+t)\,y(t)\,dt$$

2. Solve the linear Fredholm integral equation,

$$y(x)=x+\int_0^1 (x-t)\,y(t)\,dt$$

3. Solve the linear Fredholm integral equation,

$$y(x)=x+\int_0^1 (x-t)^2\,y(t)\,dt$$

4. Solve the linear Fredholm integral equation

$$y(x)=x+\lambda\int_0^1 (x-t)\,y(t)\,dt$$

5. Solve the linear Fredholm integral equation

$$y(x)=\frac{3x}{4}+\int_0^1 (xt)\,y(t)\,dt\,.$$

About the Author

Ramakanta Meher currently working as an Assistant Professor at the Department of Applied Mathematics & Humanities, Sardar Vallabhbhai National Institute of Technology. He has a teaching and research experience of above 15 years. His area of research are fluid dynamics, Linear Algebra, Differential equation and Fractional differential equation. He has published more than 55 international research publications in various reputed international journals and published 04 books. He has guided around 22 Masters students and 06 PhD students under his supervision. He has organised around 10 Short Term Training programmes/Faculty Development Programmes and delivered more than 20 invited talks in various reputed institutes.

Bei Fragen zur Produktsicherheit wenden Sie sich bitte an:
If you have any questions regarding product safety,
please contact:

Walter de Gruyter GmbH
Genthiner Straße 13
10785 Berlin
productsafety@degruyterbrill.com